MW00474864

DIGITAL DOWNFALL

DIGITAL DOWNFALL

TECHNOLOGY, CYBERATTACKS AND THE END OF THE AMERICAN REPUBLIC

HUGH TAYLOR

Copyright © 2020 by Hugh Taylor
ISBN: 978-1-7348072-2-6
Intrados Publishing
www.digitaldownfallbook.com
Cover designed by COLONFILM

CONTENTS

For my mother
Dr. Judith M. Taylor

INTRODUCTION

A Digital Fire Bell in the Night

F acing impeachment for soliciting foreign election interference in 2019, the President of the United States threatened civil war via his Twitter account.[1] What are we to make of this? Was the digital message, instantly and directly transmitted for free to a Super Bowl-sized audience, simply another in a long series of hyperbolic provocations to be ignored? Or, did it portend something more dangerous?

I have worked in the twin fields of technology and cybersecurity for twenty years. This experience led me to hear the civil war tweet as one more digital fire bell in the night, as Thomas Jefferson might have put it. It was comparable to learning about the online forum that triggered the Pittsburgh synagogue massacre or the Russian online trolls who whipped up the lethal hate at the neo-Nazi march in Charlottesville,

Virginia. The civil war tweet was another ominous signal that our society's nearly complete dependence on technology, coupled with our exposure to open global networks, is putting an increasingly fragile American republic at risk of major disruption.

I realize this sounds alarmist and paranoid, but I am not alone. A 2019 Rasmussen survey revealed two-thirds of Americans think a civil war is imminent.[2] Papers like *The Washington Post* have published opinions that Vladimir Putin is using cyberattacks and digital disinformation campaigns to turn us against each other.[3] Senate testimony by national security experts similarly warns that Russia is trying to destabilize the United States. Thought leaders on the right and left have similarly voiced concerns that we are on the verge of civil conflict. If such a massive, disruptive event occurs, it will be at our own hands—but with the aid of enemies who have the cyber motive, means and opportunity to bring us down.

What this disruption might look like is of course a matter of speculation. 2020's COVID-19 anti-lockdown protests and armed threats of civil war (the "Boogaloo")[4] that have arisen, with encouragement from online bots of unknown origin, offer some tangible clues as to what may be coming. The anti-police riots of May, 2020 in 37 cities across the US further showcase the potential for civil conflict. The disruption could involve substantial changes in Americans' civil rights and the power of their government, a loss of global standing, an economic or military disaster or an outright government collapse. Of course, I could be wrong. Indeed, I hope I'm wrong, but if I'm right, this country I love so much is in grave danger.

Being a Cassandra is a trying business. We're in a politically fraught moment, so it's tempting to read too much into a single event or risk. Trump got impeached, but his threatened civil war failed to materialize. So, was it worth the worry? Despite his failure to deliver the promised conflict, I do think the threat is worth a closer look. While no single American cyber vulnerability seems serious enough to warrant doom saying, when viewed in the aggregate, our societal risk exposure seems extreme. The civil war tweet didn't happen in a vacuum.

Trump threatened civil war in a year when the US suffered myriad foreign cyberattacks. These included Russian ransomware attacks that paralyzed cities, emergency medical systems and law enforcement—as well as massive Chinese thefts of government, military and corporate data. The year also saw Russian incursions into our power grid and industrial base.

Both Russia and China have stated they are engaged in non-shooting wars against the United States. These are "win without fighting" strategies. Russia calls its campaign "Hybrid

Warfare." Georgetown University Professor Roy Godson told the United States Senate Select Committee on Intelligence in 2017 that Russia's cyberattacks on the US, including its disinformation campaigns, were made at "the coordinated direction by the centralized authoritarian hierarchy of a combination of overt and covert techniques that propagate Russian (formerly Soviet) ideas, political military preference and undermine those of their democratic adversaries."[5] Senator Richard Burr, the Committee Chairman, further framed the issue by stating, "The American public, indeed all democratic societies, need to understand that malign actors are using old techniques with new platforms to undermine our democratic institutions."[6]

China refers to its cyber aggression against the US as "Unrestricted War." China has stolen vast amounts of corporate intellectual property in the US[7] and ransacked databases that store private information about American citizens.[8] China has also evidently hacked into US military contractors, including a brazen heist in which it stole secret codes used for submarines and other classified naval warfare data.[9]

Russia's and China's aims are not entirely clear, but they are surely not benign. Their apparent goal is to destroy America's will to fight along with any alliances that hamper their geopolitical ambitions. They want the US to step back from global commitments that inhibit their regional aggression. It's not hard to imagine that an American civil war or some comparable domestic catastrophe would be the ideal distraction to pull us out of their spheres of influence. All they need is a pretext, and many current hair-trigger geopolitical situations could easily provide one.

This book offers perspectives on the interplay between or enemies' strategies and America's digital risks. It explores how technology's accelerative effect on the global economy has transformed our society and left millions of Americans feeling economically marginalized and emotionally raw—ready for racially-motivated political manipulation. It then highlights how Russia and China have further inflamed this toxic dynamic for their own ends. This book provides insights and scenarios. It doesn't predict the future. It lays out a warning. You can decide if it's worth taking seriously.

Looking at the US today and worrying about its potential digital downfall is a Gladwellesque "blink" moment. No one thing seems so bad, but the totality of threats feels fatal. For every apparent risk, there are endless excuses and scoffing explanations. Cyber risk is bad, but it won't kill us. Russia is interfering, but it won't destroy the republic. China has stolen our trade secrets, but our economy will somehow triumph in the end. Or, it won't. One byproduct of our reliance on digital technology has been a significant distortion of our perception of reality and our critical thinking skills.

The bad thing can indeed happen, regardless of our inability to imagine it. What most expert analyses miss is the potential for disinformation and cyberattacks to work synergistically, precipitating chaos that tears American society apart. Digital and analog attack vectors, such as talk radio programs, can cross-pollinate to cause a significantly destructive event.

To take just one of many scenarios, imagine that multiple regions of the US, including any number of racial hot zones, are subject to a digital blackout, selective power outages and targeted, racially inflammatory disinformation over a period of weeks at the hands of Russian hackers. Russia is already undertaking such racially-driven campaigns as a way to meddle in the 2020 election.[10] The region's law enforcement and 911 emergency services are also shut down by Russian cyberattacks.

What do you think will happen as thousands of agitated, fearful people, who undoubtedly own many guns, run out of money, food and fuel? It's not an auspicious picture. We live in a country where people own 393 million guns,[11] but in which 47% of households don't have $400 for an emergency.[12] Perhaps a catastrophic, synergistic foreign cyberattack would result in a "spasm of violence and insurrection in this country like you've never seen," as the President's friend and former advisor, Roger Stone, warned in his own pre-impeachment tweet. Would opportunistic political leaders take advantage of such a situation to arrogate unconstitutional powers for themselves? Would anyone be able to stop them? These are questions that should be asked in this risky moment.

The "blink" sensation suggests that we are in a high state of danger... though one that's difficult to prove. That's the challenge I am undertaking in this book. I come to this task with a varied background that heightens my awareness of the risks that come from relying on technology to run nearly every sphere of American life. I have worked at Silicon Valley startups and global tech giants whose products connect the supply chains of the global industry digital fabric. For two decades, I've been immersed in the fine-grained details of America's digital ecosystem. I have a good feel for the depth of our dependence on computers. I've also seen how this system can break down.

Working in the cybersecurity field has given me a bracing sense of American cyber vulnerability. Virtually every system that runs our industries, government and military has been breached in the last few years. Our computer systems are inherently insecure, accidentally designed to contain countless security flaws. The data on every American citizen has been stolen multiple times by foreign espionage agencies, with our government only beginning to recognize the significance of this violation. Russia and China are invading our telecommunications systems, our media ecosystem and corporate systems. They've penetrated our critical infrastructure, military hardware, election systems and more. And, we don't know what we don't know. What's been publicly revealed is jarring, but we can assume the truth is far worse.

Being Jewish has also shaped my view on America's cyber-political vulnerability. The Jews have survived empires spanning ancient Egypt, Rome and Austria-Hungary, so we tend to see governments as impermanent. My father, the son of a Russian Jewish immigrant born in 1884, grew up in the Jim Crow south of the 1920s and 30s. His boyhood was spent among former slaves and Civil War veterans. In World War II, he was beaten nearly to death as a Jewish prisoner of war in Nazi Germany. His experiences influenced my thinking on racial attitudes in the United States. He always said Nazis would have caught on easily in his hometown.

My mother, as a young girl in 1939 London, watched her father try in vain to convince his Polish cousins to move to England. They ignored him and were all murdered in the Holocaust. These kinds of family memories, passed down to me as they are in so many Jewish families, have made me sensitive to risk when I see Nazis marching by torchlight in Virginia, screaming "Jews will not replace us!"

I'm also old enough to remember America when it was an analog society. My early Gen-X cohort didn't use computers, for the most part, until college or even afterwards. My MBA class received the first 900 IBM laptop computers ever made. My cohort is the last group of Americans to remember an era when people communicated on paper or over a landline phone and news came from terrestrial radio and television broadcasts. It's not that everything was so great back then, but there were more authoritative sources of the truth. The country, and its politics, were a little more grounded in reality when people weren't hovering over mobile devices during all their waking hours.

Before tech, I worked in primetime television. This experience also gave me a perspective on the current media environment. My job involved dramatizing true stories for the ABC Sunday Night Movie, working within the boundaries between fact and fiction. I then lived through the reality TV revolution, watching network standards for vulgar and exploitive content drop as competition heated up from cable.

I was even present at the moment Donald Trump made his pivot from real estate to entertainment. In 1990, my boss, Edgar Scherick, a powerful TV producer and former ABC president, struck a deal with Trump to produce "Trump Tower," a primetime soap opera—touted as the Algonquin Round Table meets "Dynasty"—with Donald Trump himself lurking in the background, a Machiavellian puller of strings in the main characters' lives. We hired Claire Labine, the co-creator of *Ryan's Hope*, to write the series bible for Trump Tower at NBC. It didn't get on the air, but in retrospect, it was a harbinger of things to come.

Which brings me to the President. While the Trump presidency has been the impetus for this project, it is not all about him. America's risky dependence on digital technology, our cyber vulnerabilities and the global impact of computing all began well before Donald Trump even thought of running for president. They will persist after his term ends. The digital wars we are fighting against Russia and China predate Trump and would have proceeded had he lost in 2016. If Hilary Clinton were president now, we would be in the midst of ongoing Chinese theft of our digital wealth and Russian cyber meddling in her presidency.

That said, Trump has uncovered the digital fault lines in American society. His unprecedented use of direct digital communication channels to speak to a large group of Americans has accelerated the risks of a breakdown in the constitutional order. If he were to see this book, he would likely call it "fake news," but he's the one who has said, on 27 separate occasions, that he will not abide by the results of an election he loses.[13] He evidently would like for these remarks to be considered "jokes."

Given how criticizing Trump seems to invite digital death threats in the new America, I've endeavored to be circumspect about how I discuss the President. I have kept comments on his personal style and policy decisions to a minimum. I don't think they're relevant to the more important matters at hand. Rather, I have criticized him and his administration for hateful messaging, demonstrable lying, perceived disloyalty to the United States and moves that undercut the constitutional order. In this regard, I think he, or any elected official who espouses such views, is deserving of rebuke. In other places, I have tried to acknowledge where he has bolstered the cyber defense of the US.

While I think the Trump administration is deserving of criticism in certain areas, I strongly believe that this issue must be addressed in a non-partisan manner. To put it into partisan terms will be a counterproductive and self-destructive experience. The impact of the threats discussed here affects everyone.

Our enemies are not interested in helping one party or another. They only want chaos and collapse. Right now, Russia appears to favor Donald Trump and the Republican Party. That is only because it suits their needs at the moment. They are happy to goose both sides of the political spectrum with their shenanigans and enjoy the resulting fights. If Russia determines, to take a hypothetical example, that a Muslim

US president and 67 non-white Senators are what will push the US over the edge into a violent crisis, they'll try to make that happen.

If and when the chaos comes, however, we may not know its exact origin. One of Trump's greatest, though perhaps inadvertent, contributions to the dialogue about American cyber insecurity is to reveal how little we truly know about what's going on. The murky provenance of Trump's civil war threats hints at the level of trouble we're facing. He appears to have been sharing a talking point that originated with Russian trolls.[14] This is the reality of the USA in 2020. Our president threatens a civil war, and we don't know for sure where he got that idea.

We are operating in the dark. Does Trump have millions of loyal followers on social media, or are they Russian bots? We don't know. Did the Russian government shut down hospitals and police departments with ransomware, or was it Russian criminal gangs with no connection to Vladimir Putin? We don't know. These are just two out of thousands of threats we cannot accurately assess because we don't know who is actually on American networks.

The fact that we don't know who is doing what in American cyber space is in itself a massive national security problem. Foreign actors can enter the US, digitally, over the Internet, with complete impunity. Our physical borders may be guarded, but our digital borders are wide open. Foreign agents can steal information, disrupt systems and engage in brazen propaganda attacks on our government and we have no clear idea who they are or what, exactly, they are up to. In this sense, Trump's antics have served a valuable purpose. They're throwing our cyber ignorance and risk exposure into relief for all to see.

Assumptions about our level of preparedness are overconfident. Though well-intended, much of the government's analysis of the problem relies on assumptions that its infrastructure and communication systems will remain functioning in a cyber crisis. Military experts overstate the cyber-readiness of the US military. Our leaders and go-to experts seem to believe the social fabric will remain intact regardless of the intensity of the attack. In my view, these are naïve perspectives given what we know about our enemies, their capabilities and agendas—coupled with the toxic narratives now dominating domestic political life.

A confluence of events in March, 2020 further reveals the potential for disaster. On March 10, the Congress released the bipartisan Cyberspace Solarium Commission report, which warned that "For over 20 years, nation-states and non-state actors have used cyberspace to subvert American power, American security and the American way of life."[15] The 122-page report, which received well-deserved praise for thoroughness and frankness, made several recommendations for how the government can and should improve its ability to defend the US from digital attacks. These included calls for strategic cyber deterrence along with giving new powers to agencies such as the Cybersecurity and Infrastructure Security Agency (CISA).

The seriousness and mature planning process envisioned by the Cyberspace Solarium Commission report, which interviewed 300 experts to reach its conclusions, stood in stark contrast to the parallel events of the government's inaction and confusion about the Coronavirus pandemic. While the Commission suggested a well-reasoned set of laws, strategies and government structures to mitigate major cyber risks, the real-life President of the United States was busy sending out contradictory, false and arguably racist messages about the pandemic. The Coronavirus response was further compounded by short-sighted earlier decisions to fire disease experts and disband the very teams that would have led the pandemic response efforts.

Furthermore, the suggestions presented in the report, which assume the highest levels of professionalism and responsibility in government, were laughably at odds with the Trump administration's decision to fire the Director of National Intelligence (DNI) weeks earlier merely for suggesting that Russia was interfering in the 2020 election.

The week before the report came out, the new, Acting DNI refused to testify before Congress about suspected Russian cyber interference in the election.[16]

The sub-optimal response to the Coronavirus and politically-driven manipulation of the nation's intelligence services show how ill-prepared the government actually is for a cyber crisis. And, with the pandemic, we at least have access to all the digital tools we need to communicate and coordinate government and civilian processes. Imagine what a national lock down would look like if we had no phones, Internet, electricity, hospitals or law enforcement. Then, imagine that Russia is deliberately injecting disinformation into the situation to amplify the fear, as it is accused of doing in Europe.[17] The likely resulting chaos is more like what we can expect from a cyber crisis, not the sober, "can-do" attitude of the well-intentioned but naïve Solarium Commission.

The Contents and Style of This Book

This book lays out patterns, facts and events that suggest the broader contours of our cyber vulnerabilities and ongoing digital wars. It explores little-known and poorly-understood threats and vulnerabilities. The goal is to illustrate the political vulnerability of the US to digital threats that may not have been considered before. It will have been a success if it raises your awareness of cyber-political risks.

The book digs into the underlying technological weaknesses in American society, business and government. I am a blogger, so consider each chapter a long blog post—self-contained but thematically linked to the others. The first two chapters deal with the nature of our technological society and its extensive vulnerabilities to cyberattacks. The chapters "Losing without Fighting," "Russia's Digital Dezinformatsiya" and "China's Unrestricted Plunder" place these weaknesses in the context of digital hostilities now affecting America's national security and political stability.

None of these threats and geopolitical dynamics are new, but as I discuss in the chapter, "Nothing's New, Except Everything," digital technology has upset the nature of American politics and foreign affairs in ways that we are only beginning to understand. This includes the underappreciated cyber risks facing the US military. The chapter "The Risk of a 'Cyber 1914'" delves into our military's cyber weaknesses.

The military might seem tangential in a discussion about the domestic political impact of cyberattacks, but a significant portion of our deterrent strategy and planned responses are based on the military's cyber capabilities. The armed forces' cyber vulnerabilities could also translate into a serious domestic political crisis if the US suffers a cyber-based defeat in an armed conflict.

The book then delves into how serious these risks might actually be. In "Who Will Not Replace Us?" I explore scenarios where foreign disinformation and cyberattacks could trigger a transformative national political event—an end to the American Republic as we know it. This chapter delves into some touchy topics. Polite Americans, such as those who create laws and regulations, dislike speaking about racism and social class resentments. If you ask a government CISO for his or her opinions on Russia's inflaming of racial tensions, they'll usually say something along the lines of "Let's not go there." To reveal the grave danger this country faces, however, we're going to have to "go there," uncomfortable as it might be.

Why? Because Russia is certainly delighted to "go there." It don't have to attend awkward Thanksgiving dinners with people who don't share their views. It's thrilled it can exploit our open networks and social media platforms to incite black people to protest mistreatment by white police officers[18] while simultaneously stoking white racial fears and resentments.[19]

Russians and Chinese security forces haven't had to swallow and defend an origin myth about American equality and exceptionalism. They can see America for what it often is: a fragile patchwork of ethnic groups that hate each other, sometimes violently. Does this last sentence make you squirm? It should. But, if we want to understand the risks facing the United States as a country, we have to confront this troubling reality.

I take the issue further in the chapter titled "Defending the Quantum States of America," which deals with our society's moral duality and what it will take to confront our problems head-on. The US is at once a global beacon of freedom and a society reeling from its bitter history of racism—a paradox that we will need to resolve if we are to survive this digital onslaught from abroad. In "Making the Machines (and Ourselves) Smarter," I look at measures we can take to address the underlying technological problems that are causing our currently high level of risk exposure.

Coping with the Inter-Disciplinary Challenge of Cybersecurity

Writing this book has forced me to confront the limits of my knowledge. If you accept that digital technology underpins virtually every element of modern life, then you will need an unattainable depth and breadth of expertise to understand how its vulnerabilities can affect our society. Getting this situation right means having a healthy command of technology, security and politics. To identify our enemies' strategies, it is necessary to have a firm grasp of history and political philosophy. One needs to know a great deal about military strategy, foreign policy and geopolitics as well.

No single individual that I can think of has all of this knowledge in equally generous parts. I certainly don't. Yet, I believe I have the kind of generalized, integrative background necessary for the task. Facing the limits of my knowledge has forced me to contend with two analytical balancing acts. First, in areas where I lack primary expertise, I rely on the works of others. I sometimes quote them at length. I feel this is appropriate because in areas like American politics, I think it's best to learn directly from the experts. In certain places, the book will therefore resemble a blogger's curated text.

The other balancing act relates to the depth of analysis on any particular issue. For each subject area discussed in the book, there is an entire library of policy literature and a long bench of experts to consult. It's easy to get pulled into bottomless, "wonky" dialogues that are distracting from the important work of identifying the country's essential digital risks.

One could write a dozen books on the topics covered in each chapter of this book. Indeed, I reference many excellent books throughout this work. These include *The Shadow War: Inside Russia's and China's Secret Operations to Defeat America*, by the CNN anchor Jim Sciutto and *Stealth War*, by Gen. Robert Spalding. The latter offers an in-depth account of

China's aggressive, but largely hidden war agenda against the US.

I have attempted to create a balance between detail and overview. This means avoiding extensive discussion of hacking and computer engineering, two subject areas that tend to dominate the discussions on cybersecurity. I feel that just as automotive engineers may not be the right authorities to consult on highway design or car insurance, so too computer scientists are not always the right people to speak to about the political impact of computers. At the same time, most political experts are unable to speak with much confidence on digital issues. Arriving at the truth and wise policy ideas can thus be a messy, frustrating experience.

The Challenge of Attribution and Certainty

Writing about foreign cyberattacks involves a process of informed speculation. There's a paucity of verifiable facts, but by examining context and professional opinion, it is possible to arrive at a reasonably confident conclusion. For example, the "Guccifer 2" Internet persona, which leaked stolen emails from the Democratic National Committee to WikiLeaks, was determined to be associated with, if not an actual branch of the Russian intelligence service. This was a guess. It was a very well-educated guess, based on an analysis performed by the best experts in cyber forensics in the American intelligence community, but it was a guess, nonetheless.

Thus, it is impossible to identify the actors behind cyberattacks with 100% certainty. Cyber attackers like to obfuscate their identities and national origins. Our open digital networks and open standards-based Internet and open computing platforms make this possible. For example, an attacker from North Korea can mask his location and make it seem as if the US is being attacked by Thailand, as was the case with the Sony Pictures breach in 2014.[20]

Figuring out America's risks in this new kind of warfare will involve some educated guesses. This is not a problem, though there is a tendency for people to want to talk about cyber defense as if the matter were on trial in court. Talking to various government people, you often find yourself getting into evidentiary game playing. For example, it's obvious from photographs and published reports[21] that China stole the designs for the F-35 fighter. Yet, if you ask someone who works in the government, they will tell you that China "only" stole unclassified information about the project and that China's knockoff fighter isn't very good.

A Note on the Technologies Discussed in This Book

At various places, this book may get into some depth discussing computer hardware, software, networks and tech standards. I will try to keep this to a minimum. However, in some cases, it is necessary to describe how systems work in order to reveal their vulnerabilities. For example, the American preference for endlessly reprogrammable PCs and phones—so-called "Turing Machines"—exposes these devices to malware. If we cannot parse the implications of the Turing Machine, then it will be difficult to understand the more easily defended alternatives.

Reasons to be Hopeful

Let's not despair. In this book, I have tried to point out reasons to be hopeful about America's potential to respond to the crisis. Some of the smartest and most capable Americans are addressing themselves to the crisis. Our underlying technology can be made more secure, if that is a priority for us as a society. The country has faced grave threats in the past and proved itself to be capable and resilient. It won't happen on its own, though. It will take focus and effort. The digital fire bell is tolling in the night. Shall we heed its call or go back to sleep? The choice is ours. The time is now.

A brief comment on the impact of the COVID-19 pandemic: This book went to press right as the COVID-19 pandemic was sweeping through the world. The event is so momentous that it works to make the concerns raised in this book seem less urgent. If anything, however, the pandemic and the reaction of the American government—and some American people—reveal the fragility of the American Republic in the face of a serious crisis. Indeed, many of the warnings continued in this book are already coming to pass, such as foreign entities digitally triggering domestic unrest in reaction to the stay-at-home orders.[22] Only time will tell if the pandemic will contribute to the digitally-driven political instability envisioned within these pages.

ACKNOWLEDGEMENTS

Writing a book of this kind requires the efforts of more than one person. I want to express my gratitude to my editors, Charlie Serabian and Ana Joldes, who helped immensely with getting the manuscript in order. The cover, designed by COLONFILM, provides a great visualization of the concepts I want to express in the book. I am grateful to Peter Miller, the Literary Lion, for encouraging me to move this project forward. I am also indebted to my wife, who had to put up with my paranoid rantings on the various topics contained in this book for months on end.

CHAPTER

1

Smart Machines, with Built-In Stupidity

This book is intended for the general reader, so I will not go into great depth on the technologies and risk factors affecting American society and politics. In fact, if you're already an expert, you can skip this chapter altogether. If you stick with it, you'll find that this short chapter examines both the origins of today's digital security flaws along with the major cyber threat categories.

I got an alarming email from my CPA in early 2019. Their firm's storage array had been hacked. All of my most personal financial information and identifying data had been stolen. There was nothing I could do about it. There was nothing the CPA firm could do about it. This was not the first time I had gotten such a notice. It wasn't even the fifth time, as I get notices like this almost every month.

How did we get here? The United States is in the midst of a cybersecurity crisis. In the first half of 2019, data breaches exposed over 4 billion records.[23] Sixty-two percent of businesses were struck by phishing or social engineering attacks in 2018.[24] According to CNN, 140 local American governments, police departments and hospitals were held hostage by ransomware in the first 10 months of 2019.[25] These are just a few of the many devastating examples of a nation that runs on digital technology that cannot be trusted.

Understanding America's cyber risk exposure requires a good sense of how digital technology works and how it can be abused. Part of this means gaining insights into how the "smart machines" that define our era are, at their core, pretty dumb. Brilliant as their designs may be from a functional perspective, modern computers and digital devices are remarkably deficient when it comes to security.

Vulnerability is designed and integrated into the core of most computer products and software programs, usually by mistake, happenstance or because of well-intentioned or misguided commitments to openness. At the same time, corporate greed and laziness further contribute to security weaknesses in products that are already not secure at the outset.

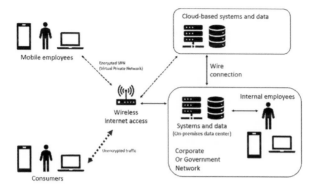

Figure 1 - Simplified reference architecture for consumer and employee access to corporate or government networks, as well as cloud-based digital assets

The Modern Digital Landscape

The technology we carry in our pockets is just one part of a much bigger digital ecosystem. Our phones and PCs connect us to millions of websites, e-commerce businesses, cloud-based digital data repositories, corporate and government networks. Figure 1 offers a highly simplified reference architecture for this environment.

Phones and PCs are known as "endpoints." The digital assets they access are housed either "on-premises" in privately-owned data centers or on cloud infrastructure, like the kind provided by Microsoft Azure or Amazon Web Services (AWS). Each endpoint, network connection, network perimeter, data repository, software application and operating system contained in this digital landscape is potentially vulnerable to hackers. Cybersecurity is the practice of defending all of these digital assets against malicious or unauthorized access.

The Tech Accidents that Led to Today's Cyber Insecurity

To appreciate our current cyber risks, it's essential to grasp the profound, built-in weaknesses of the technology we rely on to run the world. In 2020, I asked a panel of 46 industry experts if they thought our everyday technologies were designed with innate security flaws. Eighty-two percent of them ranked inherent security problems as either a major or extremely serious factor affecting security.

To understand what this means, and why this state of affairs exists, it's necessary to become familiar with the series of technological accidents that have led us to this moment in time. As with the other topics in this chapter, entire books have been written about this story, but a brief overview should suffice for our purposes.

Computers came into existence during World War II, where they were used to break enemy codes and run calculations for the nascent atomic bomb project. By the 1950s, companies like IBM had successfully marketed mainframe computers for corporate and government use. Massive and costly, they were restricted to the wealthiest and most important users.

The 1960s saw the introduction of the "minicomputer," a smaller but still quite expensive version of the mainframe. The Digital Equipment Corp (DEC) VAX and Data General Nova were examples of this technology. Mainframes and minis are still in use today, a fact that highlights the long lives of technology innovations. They have security weaknesses, but their small numbers, limited connectivity and tightly controlled management have made these problems less pressing than the ones we face today with PCs and mobile devices.

In the 1970s, hobbyists started to tinker with commercially available microprocessors and build "microcomputers." These were small and cheap enough to be used by people in their homes or small businesses. The Apple I is perhaps the most famous of these, built out of plywood in a garage by Steve Wozniak and commercialized by Steve Jobs.

The unanticipated explosion of a toy technology

The successor to the Apple I, the Apple II, was a huge hit, an industry-spawning breakout sensation that sold millions of units along with other early personal computers like Osborne, Commodore and Coleco. Impressive as they were, the early

PCs were essentially toys. They were created by enthusiasts, some of whom belonged to a hippie counterculture that valued freedom and sharing between trusted friends. In general, anyone could install any software on them at any time without having to deal with restrictive system controls.

By the late 1970s, businesses had realized the profit potential inherent in owning tiny, affordable computers. IBM, then the world's largest and most important computer company, set out to build its own personal computer. The IBM PC, which debuted in 1981, was a copy of the hobbyist models. It had an open architecture, comprising non-exclusive, mass-produced components like Intel microprocessors and a Microsoft (i.e. a microcomputer software) operating system.

Within a few years, commercial clones of the IBM PC became available. This set the stage for explosive growth in homogenous PCs worldwide. In 1983, there were 2 million PCs in the US. By 1990, that number had reached 54 million.[26] In 2014, 2 billion PCs were in use worldwide.[27] Eighty-nine percent of American homes own a computer.[28] Nine in ten Americans use the Internet, and nearly eight in ten own a smartphone.[29] The world in total now owns over 3 billion smartphones.[30] High-speed internet service, much of it wireless, is available throughout the United States.

The computer servers used to power big businesses and online services are comparable or even identical to the regular

PCs we use every day. They're bigger and more powerful, but still toy-like in their fundamental design. Unless otherwise configured, pretty much anyone can install any software on them at any time—a root cause of cyber risk.

The PC and its predecessors are "Turing Machines," meaning they can be reprogrammed endlessly to perform new tasks. This name is a reference to Alan Turing, the British mathematical genius who is credited with having many of the insights that led to the development of the first programmable computers. (He committed suicide by eating a poisoned apple, which is the inspiration for Apple Computer's name and iconography.) The Turing Machine, while revolutionary, has an inherent security flaw: bad people can reprogram it for their own purposes. This is the essence of hacking—taking control of a machine and adding malicious instruction code to make it do things it wasn't meant to do.

Security was not a high priority in the design of the early PCs. Indeed, Microsoft didn't take security seriously until the late 1990s. Until then, Microsoft and other PC makers didn't want any barriers to growth. This meant allowing virtually unlimited and restriction-free installation of software onto DOS and then Windows PCs—a great move for innovation and expansion, but a bad move for security. Viruses proliferated.

By the time hackers made a mockery of Microsoft's insecure operating system and applications as a way to pressure the company into being more serious about security,[31] it was too late. Hundreds of millions of souped-up toy computers were entrenched in nearly every area of business, government and consumer life.

The problems with open-source code and the design of the Internet

Two other technology developments compound the innate insecurity of the PC design. The first was the emergence of open-source software. While a boon to software innovation and industry growth, open-source software creates risk exposure in at least two ways. It's universally available in its source code form, so hackers can spot exploitable weaknesses right in the code.

Hackers can also create malicious versions of open-source programs and freely publish them to the world. As a result, software developers can inadvertently install malware in their programs just be using publicly available source code. There is no central governing body controlling open-source software. This is a result of the public-minded (i.e. hippie) spirit of the open-source movement. Again, this is wonderful for innovation but terrible from a security perspective.

It's unfair and a bit simplistic to argue that open-source code is less secure than proprietary code. After all, with open source, a worldwide community of experts can see exactly what's in the code, so they can identify bugs out in the open and fix them. Proprietary software, like Windows, is opaque. Its flaws are hidden, but become apparent as hackers discover exploits.

A look at today's mobile operating systems, however, reveals how proprietary code, managed the right way, can lead to better security outcomes. Industry reporting holds that Apple's proprietary iOS, which powers iPhones and iPads, is more secure than Google's open-source Android operating system. This goes against the grain, I realize. Apple is the perennial "bad guy" for making its systems closed and proprietary. But, the security data speaks for itself.

According to a 2019 article in *PC Magazine*, "Apple is usually touted as the clear winner in terms of mobile security. Apple's unprecedented control of the iPhone and iOS experience has meant that most people receive and install software updates and security fixes. That's critical, and it's a major differentiator from Android. Apple has managed to keep a tight grip over its hardware supply chain and also, through the App Store vetting process, kept control of apps created by independent developers."[32]

CNET offered a similar analysis, noting that what users tend to complain about is iOS' lack of customization, "Apple's highly patrolled walled garden has also ensured iPhone users largely stay ahead of malware without having to think about it."[33] Android security is less regulated, with phone companies issuing updates to their own versions of Android. This practice can lead to out-of-date, and therefore less secure, devices. Android is catching up, however, with automated Android updates soon to be a feature of the technology.

The other structural factor exacerbating the insecurity of digital technology has to do with the unintended consequences of the Internet's design. The Internet as we know it was originally developed for use by US Department of Defense (DoD) researchers. It was called the ARPANET, built by the DoD's Advanced Research Projects Agency (ARPA). The ARPANET was a packet-switching network, the first to use the TCP/IP protocol suite, which has since been adopted for virtually all PCs and Internet-connected devices.

The engineers who created the ARPANET, which later became the Internet, were part of a small community of academics and government researchers who knew and trusted each other. They didn't consider security to be a problem, or at least not one they couldn't solve.[34] When the Internet was made public in the 1990s, it was completely insecure—a problem that has never been adequately addressed. It allows anonymous users to traverse the globe freely. Jason Kent of Cequence Security made a marvelous comment on this topic, remarking, "I've heard it said that most of the Internet is someone's abandoned master's degree project."

The impact of corporate practices

Corporate practices further aggravate insecurity. The tech industry tends toward monopolistic behavior, with large, powerful companies resisting changes that could increase customer security. The legal foundations of tech are part of the problem. With commercial software, for example, you are not actually buying the product, so you can't sue for product defects. Rather, you're just licensing it. This is a legal loophole, enforced through a restrictive contract that indemnifies the software maker from liability. If you bought a car or a toaster as defective as software, you could sue the manufacturer out of existence. With software, you're out of luck. Companies like Microsoft can never be sued or prosecuted under product liability laws for security failures. Open-source software has a comparable flaw. For users of pure open-source code, there is no one to sue. Publishers of customized editions of open-source code hide behind licensing agreements.

Unintended consequences and the paradox of digital technology

Our digital world, a place where computers either run or affect nearly every aspect of our lives, is built on an insecure network that connects billions of toy computers invented by hippies who believe that all data, software and computer designs should be freely available to anyone in the universe—running software whose makers cannot be sued for security flaws.

This is a slight exaggeration, but any discussion of cybersecurity and the vulnerability of the United States to foreign cyber interference must proceed from this basic set of facts. It's a great illustration of unintended consequences. As characterized by the sociologist Robert Merton, unintended consequences can take the forms of unexpected benefits, unexpected drawbacks or perverse results. We see all three of these in the digital revolution that's unfolded over the last 40 years in the United States.

The digital innovations with the greatest benefit also produce perverse results and unexpected drawbacks. Facebook, for example, which was conceived as a way to connect people all over the world—and which has succeeded in this goal beyond anyone's imagination—has also opened the door to all manner of hate messages and political fraud. The Internet itself was envisioned as a way to make information universally available, among other things. It has achieved this goal, but devolved into a criminal cesspool and secret pathway for spies and warfare.

Open-source software has a similarly double-edged history. It's revolutionized software development and the entire face of the technology world. At the same time, open-source code is, by its very nature, highly insecure.

The paradox of all this is that there wouldn't be an Internet or Facebook, or even the PC revolution itself, if the technology hadn't been so flawed to begin with. The very toy-like nature of it, the slovenly non-design of so many of its components, enabled its explosive uptake. If a single corporation had tried to make PCs and the Internet happen, we would probably not have 2 billion PCs and 3 billion smartphones in connected operation today.

Conclusion

A great deal of our cyber risk arises due to inherent design flaws in computers and smartphones. These flaws got built into our technology because of a series of accidents and questionable decisions that are now yielding unintended consequences. Many serious cyber threats have evolved to exploit these weaknesses.

CHAPTER
2

The (non) Impact of 20 Years of Cyber Warnings

I t's almost funny. In October of 2018, the General Accounting Office (GAO) prepared a report on Weapons System Cybersecurity[35] at the behest of the US Senate's Committee on Armed Services. The news was not good. The report's subtitle, "DOD Just Beginning to Grapple with Scale of Vulnerabilities," reads like a joke, considering that the Department of Defense (DOD) has been trying to get out ahead of cyber vulnerabilities since at least 1998.[36] It's easy to be snarky about this, so I won't, but seriously?

The GAO report is a recent, but hardly singular example of how the inherent cyber vulnerabilities described in the previous chapter manifest in crisis-level risk exposure in the US government. The stakes are high. In financial terms, the report notes, "DOD plans to spend about $1.66 trillion to develop its current portfolio of major weapon systems."

Money is the least of it, though. According to the report: "In operational testing, DOD routinely found mission-critical cyber vulnerabilities in systems that were under development, yet program officials GAO met with believed their systems were secure and discounted some test results as unrealistic. Using relatively simple tools and techniques, testers were able to take control of systems and largely operate undetected, due in part to basic issues such as poor password management and unencrypted communications."

In one case, testers were able to guess the password to a weapon's system in less than 10 seconds. "Even 'air gapped' systems that do not directly connect to the Internet for security reasons could potentially be accessed by other means, such as USB devices and compact discs," the report stated, then added, "Weapon systems have a wide variety of interfaces, some of which are not obvious, that could be used as pathways for adversaries to access the systems."

Then, in June, 2019, the Senate Permanent Subcommittee on Investigations published a 99-page report titled *Federal Cybersecurity: America's Data at Risk.*[37] The committee, chaired by Senators Rob Portman of Ohio and Tom Carper of

West Virginia, revealed some shocking discoveries about lax protections of citizen data in seven federal agencies, including the Department of Transportation, DHS, the Department of Education and the Social Security Administration. The report was the result of a review of 10 years of Inspector General reports.

The report reviewed a wide range of alarming government security deficiencies. Problems include reliance on outdated, unsupported systems, failure to apply mandatory security patches, neglecting to keep track of hardware and software and more. The report also highlights an increase in cyber incidents, which jumped from 5,500 in 2006 to 77,000 in 2015. (The number dropped to 35,277 cyber incidents in 2017, but this reflects a change in definitions, rather than a decline in hacking activity.)

Examples of security weaknesses include the use of unsupported Windows XP and Windows Server 2003 at DHS and decades-old legacy systems at the Transportation Department and the Social Security Administration. These agencies store personal information about American citizens and sensitive government data on these vulnerable systems.

The revelations in the Federal Cybersecurity Report are also disappointing when one considers that the issue has been the focus of administrations going back to President Bill Clinton, at least in theory. In 1999, Clinton announced a $1.46 billion initiative to improve government computer security. The aim was to protect against terrorist attacks that might target the nation's infrastructure, such as power plants, telecommunications, banking, transportation and emergency services.[38] This led to the creation of the National Infrastructure Protection Center (NIPC) and the Federal Computer Incident Response Center (FERC) along with a variety of other measures. The Clinton program, and comparable initiatives under Bush and Obama, were sensible, necessary moves. As we know from the events of the last decade, however, they were not sufficient to deliver effective cyber defense to the US.

So little seems to have been achieved in 20 years. The underlying design flaws in our technology are part of the problem. To further understand America's cyber crisis, it's necessary to have insights into how hackers penetrate networks and information systems.

Understanding how Hackers Succeed in Disrupting Digital Systems

The term "hacking" is a bit of slang with a variety of etymological antecedents. It generally refers to computer programmers "hacking away" at code as if they were wielding axes against logs. Hacking is about making a piece of technology do something it was not intended to do. The concept goes back to the earliest days of computing, with naughty MIT students picking the locks to the computer lab in the 1950s so they could code all night.[39]

Today, the word has a darker connotation. Hackers are people who discover exploitable weaknesses in computer systems to disrupt their functioning, eavesdrop, steal data and so on. Some of this is a pure nuisance, intended to show off technical prowess in the face of poor engineering at the corporate level. Indeed, hackers are something of a tribe, often committed to anarchistic, anti-corporate or anti-capitalist political philosophies.

Politically-motivated nuisance, crime, warfare or all three?

Hacking has since expanded to become a criminal enterprise, a form of warfare or both. Malicious hackers, known as "Black Hats," like the bad guys in old Westerns, break into computer systems to breach data, commit fraud, conduct ransomware attacks, use the victim's CPU to mine for cryptocurrencies and so forth. "White Hats" are professional hackers (often former Black Hats) who help corporations and governments discover their cyber weaknesses.

To understand how hacking works, we need to digress briefly into how computers function. Every digital device, from PCs to smartphones to Internet of Things (IoT) devices, has a comparable design. A microprocessor serves as the "brain" of the device. It takes data inputs from keyboards and touch screens, processes it and outputs the results to screens or printers or sends it somewhere else on the network.

Controlling the microprocessor is the Operating System software, or OS. This could be Windows, Android, iOS, Linux, Unix or others. The OS software is what makes the computer function. The OS can be configured to allow other software to be installed on the machine. Application software runs on top of the OS. It "applies" the computer to a task, like sorting data or managing accounting functions. Much of the time, the application software connects to a database program.

At the hardware level, a special kind of software known as "firmware" controls how individual electronic elements of the system work. There is firmware that controls the motherboard, the disk drive, the keyboard and so on. Firmware is hard-coded into the circuitry itself at the factory, though it can usually be updated by system administrators.

Each component in this design scheme is vulnerable to hackers. Hackers can take over the OS and make the computer do all sorts of unintended things like exporting data from a database or just switching itself off. Databases can be hacked, stolen or corrupted. Firmware can have malware inserted into it, causing the machine to malfunction or shut down. Application software can be commandeered as well.

How hackers penetrate networks and information systems

How do hackers get into the system? For a computer that's undefended, the process is remarkably easy. A hacker can log on and install malicious software. Today, however, computers have layers of protection around them. To succeed in penetrating a system today, the hacker must have what's known as an "exploit." An exploit is a technique that will allow a malicious actor to take up residence on the computer system.

Typically, an exploit is a gap in secure functionality that was created by accident. Software developers and system architects work at a fast clip, and usually in groups. A lot of the time, more than one group is responsible for the design and production of a computer system. This results in bugs, which are mistakes in the code that allow a malicious actor to make the software break down and do something it was not designed to do. According to industry experts like author Steve McConnell, the average software program will have between 15 to 50 errors for every thousand lines of code.[40] Thus, a 20,000-line iOS app could contain over 300 bugs. Hackers are good at discovering these mistakes and exploiting them.

A cyberattack follows a sequence of steps known as the "attack chain."[41] (A cybersecurity team works in reverse, in a "kill chain" sequence.) The typical attack chain starts with reconnaissance, where the hacker picks a target and looks for vulnerabilities. This is followed by the development of malware, or "weaponization" and then delivery, where the hacker implants the malware in the target's systems. Delivery is often performed via email phishing attacks.

After delivery, the hacker installs the malware. Or, the malware installs itself, if it is so instructed. With malware installed, the attacker assumes "Command and Control," or persistent access to the target's networks and digital assets. Finally, the hacker is able to perform his or her actions, e.g. data corruption, systemic vandalism, espionage and so forth. The Turing Machine design makes it relatively easy for hackers to execute the attack chain, reprogramming the machine to do what they want.

Structural problems and lax security policies

A complicating factor here is the heterogeneous and fragmented nature of software. When we start up Windows or application software, it seems as if a single software entity is operating on the machine. In reality, a program like Windows consists of hundreds of smaller code components, operating together to create the Windows user experience. These components are built by different teams around the world. It's remarkably easy for this process to yield numerous exploitable vulnerabilities.

Over time, the software maker and (hopefully) friendly White Hat hackers discover the exploits and fix them by issuing software "patches." A patch is a piece of code that remediates a vulnerability. Companies like Microsoft issue patches regularly. This is why iOS, with its rigorous and automated system updates, is considered more secure than Android, with its less predictable updating.

System updates, which include patches, translate into improved security. The process of "patch management" is essential to maintaining a strong security posture. The opposite is also true. Lax or sluggish patch management means poor security. Unpatched systems contain exploits that leave them vulnerable to compromise. Indeed, some of the worst hacks in recent history, such as the Equifax data breach,[42] were the result of poor patch management.

The irony of cyber vulnerability is that, despite all of the advanced hacking techniques available, it is often the simplest mistakes and most routine negligence that creates the greatest risk exposure. Hackers employ automated software that scans the world for unpatched systems and other easy-to-miss configuration errors that leave systems exposed. To put the matter into perspective, *Computer Weekly* wrote that the notorious Office of Personnel Management (OPM) hack in 2015 could have been prevented if the agency had simply followed basic security procedures.[43]

The "dark web"

What do hackers do with the data they steal? The common misconception is that the hacker, who is of course wearing a hoodie and drinking Jolt cola, steals the victim's credit card number and then buys things with it at target.com. This may once have been true, but today's hackers are far more organized. They sell their stolen wares on a hard-to-reach segment of the Internet known as the "dark web."

The dark web is a scattered collection of Internet properties that can only be accessed by special software tools. Dark web sites are designed not to be indexed by Google, so regular web users can't find them by searching. People have to know where to look. The hacker underground publishes information about reaching the dark web, but one has to be a pretty accomplished hacker to even know where to look and how to use the required tools.

It's a criminal bazaar, in essence. On the dark web, fraudsters and other criminals can buy and sell identities, credit card numbers, social security numbers and corporate network logins. For instance, if a hacker wants to penetrate General Motors' networks, he or she doesn't have to do the hard work of breaking in. They can buy stolen network credentials for GM on the dark web. There are also allegedly other criminal activities taking place on the dark web, such as drug deals, but that's beyond the scope of this discussion.

Common Cyber Threat Vectors Used Against the United States

Hackers use dozens of different techniques to execute their attack chains. The following comprise the main threat vectors and hacking processes seen by cybersecurity teams in the US:

- **Email attacks, e.g. phishing and spear phishing**—these attacks involve tricking the recipient of an email into opening a file that contains malware. They then click on a link that loads malware onto the user's machine, or gets the victim to "log in" to a fake, but realistic-looking website in order to steal their log in credentials. A remarkable 94% of malware is delivered by email.[44] A variant of this attack is the "CEO fraud," wherein the attacker impersonates a company CEO to trick the email recipient into wiring funds to a bank account.

- **Malware**—malicious software (malware) is code that executes steps in the attack chain like seizing command and control, mining for cryptocurrencies (crypto-jacking) and stealing data. According to Symantec, one in thirteen web requests leads to an attempt to download malware.[45]

- **"Zero Days"**—given the large number of software bugs being produced every day, a portion of them will remain unknown to security professionals. They are not among the many digital "signatures" filtered by anti-virus software. Undetected, these bugs can be exploited without warning, i.e. with zero days' notice to virus filters. Zero days are very harmful because they can mask an intruder's entry into a system. Attackers may persist in a target system for months before being detected.

- **Advanced Persistent Threats (APTs)**—these are mostly the domain of highly sophisticated nation state actors. An APT is able to penetrate and linger in a target's systems for a long period of time, perhaps in a dormant state. These are sometimes called "implants," which can be remotely activated to perform steps in the attack chain.

- **Ransomware**—in this type of attack, the hacker penetrates the target's system and encrypts their data, demanding a ransom, usually paid in cryptocurrency to unencrypt it. Many victims pay and the attackers usually follow through with the promised decryption. It's big business according to the *MIT Technology*

Review, with an estimated $7.5 billion paid in ransoms in 2019.[46]

- **Distributed Denial of Service (DDoS) attacks**— hackers take over many machines (perhaps millions) and command them to interact with the target system. The target, e.g. a government or e-commerce site, is unable to handle millions of simultaneous requests. It shuts down or makes it impossible for anyone to use it legitimately. Industry research holds that 51% of businesses experienced denial of service attacks in 2018.[47]

- **"Grayware"**—these are legitimate software programs that users willingly download for useful purposes, e.g. driving directions or photo storage. However, the app is spying on them and grabbing a lot of unrelated personal data. The user agreement may permit this, but it's not ethical. And, the data they collect may then get hacked by a malicious third party if it's stored on insecure cloud servers.

- **Spoofing**—this is a common term, though a better word for this might be "pretending." Spoofing involves tricking a victim into thinking it is interacting with a technology when, in fact, it's a hacker in disguise. This might mean visiting a website that looks like a bank, but which is actually a "spoofed" version of the bank's site. This way, the hacker can steal the victim's bank account data. Sometimes, spoofing is machine-to-machine, e.g. GPS spoofing, where an attacker tricks a GPS receiver into wrongly orienting itself on an electronic map.

- **Firmware threats**—the firmware that powers digital hardware can also be hacked. Because firmware is acting at a lower systemic level than the OS, an unauthorized change in firmware can be almost impossible to detect. Meanwhile, compromised firmware can have nearly unlimited authority to steal data or disrupt the system. In some cases, such as with the suspected 2012 Iranian hack of the Saudi Aramco state oil company, the result was the permanent "bricking" or shutdown of 35,000 PCs.[48] A related risk is the unauthorized installation of rogue hardware on digital equipment. China was so suspected of placing small spy chips on SuperMicro servers in 2018.[49]

- **Privileged account abuse**—the administrators of IT systems are called "privileged users." They have the ability to set up, delete or reconfigure systems and user accounts. As a result, they are choice targets for hackers. If a hacker can take over a privileged user's account, he or she can wreak havoc on the target.

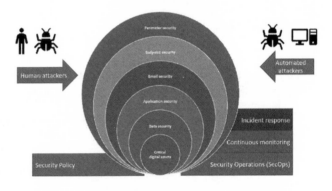

Figure 2 - Defense in depth approach to protecting critical digital assets from threats

Highlights of Cybersecurity

For each threat, there is a defense. Cybersecurity departments concern themselves with protecting digital assets from attacks in ways that align with the main threat vectors. Figure 2 depicts the kind of multi-layered "Defense in Depth" strategy undertaken by most large security teams.

Each area of the enterprise IT landscape has its own protection. There's security for the perimeter, e.g. firewalls, endpoint protection, email security, application security, data security and more. Security policies, or overarching rules and procedures govern the protection process. The Security Operations (SecOps) team uses continuous monitoring tools to check for intrusions and anomalies that might suggest an attack. An incident response teams handles alerts from the many different security tools in use.

As you might imagine, the setup shown in the figure is costly and people-intensive. The scale and scope of the defenses themselves create problems of coordination and orderly incident response. The work can be stressful, and burnout is common in SecOps teams. And, there is a significant shortage of skilled security people in the current IT labor market. *CSO Online*, a major industry site, estimates that there will be 3.5 million open cybersecurity positions unfilled by 2021.[50] It is still possible, however, with the approach shown here, to mount a reasonably strong cyber defense.

Trends that are Exacerbating American Cyber Risk

Our current digital age is one of feverish innovation and competition for digital business. Every major US business has a web presence, with many companies now actively engaged in "Digital Transformation" projects that involve using technology to engage more deeply with customers and improve their operations. Federal, state and local governments are similarly driven. They offer increasingly extensive online and e-government services.

Having this much digital technology in our lives is an invitation to cybersecurity risk. However, a number of trends are further amplifying the extant digital risk. These include Big Data and Artificial Intelligence, The Internet of Things (IoT), mobility and pervasive wireless networks, cloud computing, intelligent bots and standards-based application programming interfaces.

Big Data, Artificial Intelligence (AI) and Machine Learning (ML)

The prevalence of digital technology in American life provides the foundation for new trends that offer great promise, but also expose the country to cyber risk and related threats like invasion of privacy. For example, the "Big Data" trend, Artificial Intelligence (AI) and Machine Learning (ML) enable machines to solve problems and uncover insights that make businesses run more profitably and governments more effective at providing services to citizens.

The premise of Big Data, AI and ML is that the massive volumes of data that human activity produces every day can be mined for useful insights and patterns. A simple example might be the way Amazon recommends a book for you based on what you've already purchased. The software that runs Amazon examines your data, along with that of millions of other people and makes a suggestion.

AI gets better by "learning" how well its pattern discoveries are doing. In the Amazon example, ML keeps watching whether you click on its book suggestions. It will then "retrain" itself to do better in making book suggestions based on what it "sees."

Google search and Facebook social feeds work similarly. AI-driven data analytics also pervade business fields like stock market trading, in which software makes many buy/sell decisions. It's affecting medicine, with AI-assisted diagnostic tools assessing patient symptoms in the context of enormous patient data sets that no human doctor could ever analyze.

A number of security risks arise with Big Data, AI and ML. One problem is their reliance on having access to the data itself. Running analytics on big data sets usually means pooling data volumes and storing them somewhere, often on cloud servers that are not properly configured. Thus, we see massive data breaches that involve accidental exposure of data that was stored for analysis but left unguarded, or stored on poorly-configured and therefore insecure systems. The

Capital One breach of 2019, in which 100 million records were stolen, was an example of this problem.[51]

All this data analytics can also be harnessed for malicious or illegal purposes. It can be a tool of surveillance, which is already happening in China. Data analytics can help distort our politics, as the Cambridge Analytica scandal revealed.

AI is also enabling the production of "deepfake" videos. Deepfakes (a combination of "deep learning" and "fake") is a form of synthetic media that arises from the use of AI and video software. By studying the video of a politician, for instance, it is now possible for a machine to create a wholly synthetic video of that person appearing to be drunk, committing a crime or saying things that he or she has not said. Deepfakes can also manifest as revenge porn or just about any malicious and libelous fake news one could ever imagine.

Congressional leaders, fearing that Russia or other entities will create deepfakes to disrupt the upcoming elections, has introduced legislation to punish their use. This has resulted in opposition from organizations like the Electronic Frontier Foundation (EFF) and the *Columbia Journalism Review*, which believe that existing laws already cover the issue and worry that the proposed legislation will stifle free speech.[52]

The IoT and smart devices in the home, in industry and in "smart cities"

The IoT refers to the growing practice of connecting digital devices to the Internet—without a human user on the other end. A digital device like a thermostat is considered a "thing" in this context. There are several different IoTs, including an Industrial IoT, which mostly consists of sensors for industrial production, oil rigs and so forth. There is an agricultural IoT, which uses sensors in the soil and air to gather data about farming conditions. A consumer IoT comprises billions of "smart devices" like thermostats, doorbells and digital locks as well as toys, digital watches and drones.

Several distinct characteristics of the IoT deserve attention from a security perspective. One is pure scale. The IoT is huge and growing, with over 30 billion devices in it now, but with a forecast size expected to exceed 75 billion devices by 2025.[53] Any entity of this size and diversity is going to be impossible to secure. Add to this the fact that virtually all of the "things" in the IoT are made in China. This is problematic for two reasons. The mass-manufacturing and rapid market delivery of these low-cost products makes security a low priority for their makers. In some cases, as the US Navy discovered with Chinese-made drones, is that "things" often send data back to China, with unknown but suspicious intent.

Some of the "things" in the IoT may actually be quite large and dangerous. These include connected cars and trucks as well as equipment that is controlled by IoT devices. Oil refineries and natural gas pipelines are two highly combustible examples. The power grid also increasingly relies on IoT devices for proper functioning.[54] A cyber attacker could cause real-world injury and mayhem by hijacking such "things." According to CSO Online, 61% of organizations have experienced an IoT security incident.[55] Symantec research data reveals that IoT devices average as many as 5,200 attacks per month.[56]

The IoT also generates immense amounts of data. This data is created and is often stored at "the edge," or outside of conventional data centers or well-secured cloud infrastructure. There, it is vulnerable to breach.

The emerging "smart city" trend is largely about the IoT. The smart city uses IoT sensors and devices to improve civic life. For example, a smart city might employ Wi-Fi-connected parking meter sensors to direct traffic to open parking spots. The city could use IoT sensors to detect street lamps without bulbs, and on and on. This is all great, but it's also potentially insecure. Malicious actors, who have already demonstrated ample ability to hack municipalities,[57] might find the smart city an inviting target for mischief.

Mobile computing and pervasive wireless networks

Mobile computing is about more than just smartphones, though they are a big part of it. Many modern workers use mobile computing on a daily basis. This might include a field service technician who uses a mobile computer to complete his or her work or a salesperson who uses a corporate-issued tablet to interact with the firm's Enterprise Resource Management (ERP) and order management systems.

To function, mobile computing needs extensive wireless networks. The major cellular networks from AT&T and Verizon form the backbone of this wireless infrastructure, but there are many other comparable networks. These include corporate and municipal Wi-Fi as well as private wireless networks used by the government, military and so forth.

Mobile and wireless add to everyone's cyber vulnerability by adding millions of network "endpoints" to corporate and government infrastructure. The network perimeter, which used to offer a simple, fixed point of defense, no longer exists. The perimeter is everywhere people go with their mobile, network-connected devices.

A government worker using a government-issued smartphone on a coffee shop's public Wi-Fi is exposing the government's networks to hackers who can detect and penetrate the device in the public realm. The hacker might also be able to eavesdrop on the communications between the device and the network, stealing confidential data or network log in credentials. In these ways, hacking the government gets quite a lot easier than trying to land a phishing email inside a government agency.

The dependency on wireless networks creates a separate kind of risk, which has the potential to disrupt work by interfering with the network's functioning. If the attacker's goal is to paralyze corporate and government operations, the ability to shut down wireless networks becomes another valuable weapon in his or her arsenal.

Bots, intelligent agents and RPA

A bot is a software robot, capable of acting on its own, within a set of parameters established by the programmer. They're not new. We've all interacted with very simple bots when we announce our account numbers to an automated bank customer service line. Bots are also common on customer support chat boxes and so forth. They "listen" for us to tell them keywords. Then, based on those words, they share information they "think" we will want. A lot of the time, the bots are right, and everyone can save a lot of time and effort.

As AI gets more advanced, bots are becoming smarter. An intelligent agent is a higher-level bot that can be more creative in its response to inputs. They might be able to use rules-based algorithms to suggest solutions to subjective problems. The latest and smartest incarnation of intelligent agents is a technology known as Robotic Process Automation (RPA). In RPA, software robots can be instructed to perform simple thinking tasks previously done by human beings. For example, an RPA robot might be able to open a spreadsheet attached to an email, run calculations on it and forward the message to a human being based on the outcome of those calculations.

RPA has positive and negative potential uses, like so many other technologies. On the one hand, it can save money for companies, but it might lead to layoffs of thousands of people. RPA agents can also be programmed to do bad things. This is still something of a hypothetical, but a hacker could take over an RPA system and create robots that steal account information, for example.

Bots can also participate in social media. The Senate investigations into the 2016 election interference and the Mueller Report both cited the use of Twitter bots and bots on Facebook as influencers on the public opinion about Hillary Clinton and Donald Trump. Bots are becoming more sophisticated over time, more capable of mimicking human language and thought patterns. The Senate eventually issued a final report in 2020 that endorsed the view of the

US intelligence community that Russia had interfered in the 2016 election.

Cloud computing

Cloud computing has been sufficiently hyped that it's easy to lose track of what it actually is and why it matters. The cloud spans two different concepts that blur into a single entity in most of our minds. It's an ownership and business model, for one thing. Instead of operating computer systems in private data centers, companies are now renting computer space from cloud providers like Amazon and Microsoft. Virtualization, a technology that enables multiple systems to run on a single physical computer, is the main enabling factor here.

The cloud is also a software architecture. In the cloud, users and the software they're accessing are abstracted from one another. The software is not local. It's "in a cloud," so to speak, somewhere else. The user is not responsible for managing the digital assets. The user can "spin up" and "spin down" cloud software as needed, place data in the cloud as required and more. It's endlessly flexible. Cloud computing gets the user out of the capital investment necessary to build data centers and buy servers and so forth. The cloud provider shoulders that investment and makes money by renting out the resulting infrastructure.

The cloud is a mixed proposition in terms of security. The core cloud data centers themselves are usually pretty secure. At the very least, they employ more security people, and of a higher quality than most companies can ever dream of hiring. They have the best security countermeasures. However, the cloud providers invariably work on what's known as a "dual" or "two-tier" security model.

The cloud provider only secures the actual machines. The customer is responsible for securing his or her own software and data. This makes sense, because it would be impossible for Amazon to know, for instance, who is allowed to access one's data. But, as breaches like Capital One's have shown, accidental misconfigurations in the cloud can expose data to unauthorized access.

The 2019 hack of Wyze Labs also highlighted the security risks of cloud computing and their potential impact on consumers. In this case, the company, which manufactures affordable security cameras, had to announce that the private information of 2.4 million customers had been exposed due to a cloud breach. In their case, the cause was innocent, but revealing. A new employee set up a cloud database for use in data analytics. He or she had not been trained or informed about the company's security policies.[58] The resulting database was open to unauthorized access.

Standards-based APIs and universal system interoperability

Application Programming Interfaces (APIs) are software programs that enable one piece of software to interact with another. (APIs are also used to connect pieces of hardware.) For example, if you use your banking app to check your stock investments, most likely the banking app is using an API to reach out to your investment house and request your stock data. This is a machine-to-machine interaction.

APIs are not new, but in recent years they have been subject to a revolution in openness and universal interoperability. By using open standards (technology that no one owns, but everyone agrees on), APIs now make it possible for a modern application to inter-operate with pretty much any other software in the world.

This is at once amazing and frightening. A modern, standards-based API can pull data from a comparably configured data source. It can send and receive procedural calls, making remote software take actions on command. As you might imagine, this can be a great way for hackers to cause harm. The API standards don't provide for much built-in security, so it's a real problem. An example of this vulnerability arose when Salesforce Marketing Cloud accidentally exposed user data to external access through its API in 2018.[59]

API risk is severe, especially when APIs connect more than one company. The Ponemon Institute, which conducts surveys and studies of cybersecurity subjects, found that 61% of organizations in the US experienced a data breach caused by a third party or vendor. A further 57% were unable to know for sure if their vendors maintained adequate security policies or defenses. Fewer than half even check their vendors' policies and security practices.[60]

Conclusion

The United States is experiencing a cyber crisis, with unimaginable levels of risk affecting the digital technology that runs most of American businesses, government and life in general. Defenses have arisen in parallel, however, to defend digital assets against attacks. It's a challenging process. New trends like cloud computing and the IoT pose further risks, with technology advancing more rapidly than it can be effectively defended.

CHAPTER 3

Losing Without Fighting

Driving around LA on a Sunday morning with the radio on in 1991, I heard the announcer come on the air and say, "Moscow is in crisis as the Soviet Union ceases to be. And now, back to Beatles Brunch." Yes, what was arguably the single most important world event since August of 1945 came and went during a commercial break advertising a 1960s rock and roll nostalgia show.

It feels as if we've been pretty checked out ever since, as a people. Engrossed in exciting and addictive technologies, we carry on with our lives, unaware and uninterested in increasingly threatening world events. Over the last two decades, we've devolved into states of war with Russia and

China without seeming to notice or care very much. Each adversary treats us in their own distinct way. Russia is flashier and visibly dangerous. China is quieter but arguably more dangerous. These hostilities, which comprise a great deal of cyberwarfare, form a treacherous alignment with the excessive reliance on digital technology described in the previous chapter—as well as with the attendant cyber vulnerabilities this reliance has caused.

Can you lose a war you don't even know you're fighting? Whether we realize it or not, the United States is currently locked in serious, warlike geopolitical struggles against Russia and China. Non-violent, at least for now, these conflicts pose an existential threat to the United States as we know it. The cyber crisis described in the previous chapter serves to weaken the United States in these significant global confrontations.

For sure, Russia and China consider themselves to be at war with the US. They have devoted their most elite military units to the fight. The publicly-known results are devastating. We should assume the non-public data is far worse. These wars are based on "win without fighting" strategies, which makes perfect sense given that neither China nor Russia could currently defeat the US militarily. To try would be a lethal, wasteful exercise. They can get what they want—regional, then global domination—through technological means.

The World Since 1991

Major events can slip past us, even if we're paying attention. The headlines tell the big story, but the greater truth can remain hidden. So it was in 1991, a year of epoch-making changes, some of which were not at all apparent at the time. That year of Soviet collapse also saw the launch of the World Wide Web. The web remained an esoteric technological innovation for several more years. Few people who saw it in its early form, as I did, could have predicted the impact it would have on business, government and media in the ensuing decades. The web caught on, driving the popularity of the Internet for consumers—leading to another massive growth spurt for PC sales.

Linus Torvalds debuted the Linux open-source operating system in 1991. This, too, proved transformative. Linux has since become one of the world's dominant operating systems for the computer servers that run platforms like Facebook and Google. Tech giants like IBM revised their entire strategies to be organized around Linux.

The US also fought a war in Iraq in 1991, a war we won easily. It was the first US military conflict in the post-Cold War era. The war changed America's role in the Middle East, but the biggest and least-understood effect of the war was its impact on China.

The Chinese, who at the time felt their army was inferior to Saddam Hussein's, watched in horror as the US military destroyed the Iraqis in three weeks, suffering a few hundred casualties in the process. The Gulf War of 1991 thus spurred the government of Deng Xiaoping, China's paramount leader, to undertake a radical modernization plan for the whole country—a process that we are all very much experiencing today. Prior to that, China was considered a developing nation. As a result of the shock of 1991, China is now the second largest economy in the world.

The events of 1991 were met with a series of poor assumptions in the United States. It was hard to take China too seriously until it was too late. They were a nuclear power and all, but they were so poor and weak that few people in positions of authority considered China a global threat. Regarding Russia, American leaders believed the country would turn into a democracy because of the influence of capitalism. In geopolitical terms, Russia was also perceived as poor and weak, which was an accurate but short-sighted view.

The US in the 1990s basked in the position of being the richest and most powerful nation in the history of the planet Earth. Aside from the Gulf War and other minor military operations, the country was at peace. The Cold War was over. We had won, and no one could ever touch us!

Geopolitics does not stand still, however, and the US has still not adapted to the post-Cold-War world. As George Friedman, founder of the influential publication *Geopolitical Futures* put it, "The United States emerged from the Cold War in a state of surprise. It has never fully adapted to the post-Cold War world, and in particular to the need for a different strategy. The US currently has a problem defining what issues matter to the country, and recognizing that many, if not most, don't matter."[61]

While the US was in a state of surprise, as Friedman might say, or preoccupied with the dot.com boom, O.J., 9/11, wars in Iraq and Afghanistan, the Great Recession and a blue dress, the world changed. We were otherwise distracted, with the average American spending six hours per day online, half of which were on mobile devices, by 2018.[62] As we focused on Candy Crush and Angry Birds, Russia began to assert itself in Europe in the years following Vladimir Putin's (highly suspicious and likely illegitimate) re-election as Russian President in 2012. China's influence and global ambitions also became operational in this last decade.

Now, after decades of global change we mostly missed, we are in the midst of a global power struggle. We are waking up to expansionist strategies in Eastern Europe and the Western Pacific. In essence, the digital attack front of these so far bloodless wars are the cyberattacks and aggression we are experiencing from Russia and China today.

Russia: Seeking Eurasian Dominance and the Destruction of the US and EU

The United States is caught up in a European conflict that is largely non-violent, but does erupt into actual combat at times. The conflict has Russia pursuing a long-term strategic campaign to dominate its former Soviet republics, Central European puppet regimes and the EU. American leadership's perspective on this conflict, and Russia in general, has been, to channel George W. Bush, one of "misunderestimation" for nearly 20 years.

As Russia engaged in cyberattacks against its neighbors, poisoned Russian expatriates in England, annexed Crimea (the first such territorial grab since World War II) and invaded Ukraine while transparently lying about it, the Obama administration took a dismissive view of their plans and capabilities. As Jim Sciutto described in his book, *The Shadow War*, "At the G7 Summit in 2014, President Obama relegated Russia to 'regional power' status, saying that its territorial ambitions 'belonged in the nineteenth century.'"[63]

The 2016 election of Donald Trump, Russia's preferred candidate,[64] which featured audacious Russian interference in the American political system, has served to further obscure

Russia's attacks on the US. Trump has blocked efforts to investigate Russia's aggressive stance against the US and formulate an effective strategic response.

Fearing Trump, the US Senate has weakened legislation meant to prevent future Russian election meddling.[65] Indeed, Trump has lobbied to investigate the investigators who originally looked into Russia's hacking of the 2016 election and fired his FBI Director to stop the matter from being thoroughly examined—calling the whole affair a "hoax."[66] Trump has publicly stated that he does not believe his own intelligence service's assessments on the matter, preferring to take Vladimir Putin's word over those of American officials.[67] From there, Trump appears to have advanced Russia's interests in its war against Ukraine by withholding American military aid to that country.[68]

While the main focus of Congressional investigations and media reporting has been on Russian election interference, the broader conflict is arguably more serious. As Sciutto said in his book, "It's as if Russia and China have started a new Cold War and American didn't notice." He cited an interview with General Michael Hayden, the former CIA Director, who laid out Russia's new doctrine of "Hybrid War"—a "strategy of attacking an adversary while remaining just under the threshold of conventional war."[69]

Is Russia simply a "regional power" trying to dominate its neighbors, as Obama would have it? Or is there a bigger game afoot? The fog of hybrid war makes it difficult to establish what's going on with absolute certainty, but an analysis of the situation by experts suggests that a much larger and more threatening strategy is at work.

The motivations of the Kremlin, and Putin in particular, are complex and rooted in political philosophies of fascism and Russian history that are little known to most Americans. Timothy Snyder, the scholar and author of *On Tyranny* and *The Road to Unfreedom*, has delved deeply into this subject. He discusses the influence of a fascist political philosopher named Ivan Ilyin (1883-1954) on Putin and others in his circle. Ilyin viewed fascism as the path to the "redemption" of Russia from toxic Western forces.

So inspired, Putin views himself, according to Snyder, as Russia's "redeemer," the fascist savior of the pure Russian state. Such a pure state can never be part of any European project. Putin and his allies see the west as being morally corrupt, portraying Russia as an innocent virgin that must be defended against the "perverted" advances of a degenerate continent. Putin's anti-gay rhetoric is similarly rooted in this sexualized national symbolism that informs modern Russian fascist thinking.

In practical terms, protecting the "purity" of Russia means entrusting power to a small group of very rich oligarchs who appoint Russia's leader. This leader then governs through "managed democracy"—meaning sham elections, meaningless political parties and an essentially totalitarian regime[70] where dissent is a crime. This may start to sound familiar, for those paying attention at home: billionaires fund a US presidential candidate, take advantage of election irregularities, position him as the "savior" of "real Americans" and push for criminalizing dissent, e.g. making news reporters susceptible to harsh libel laws.

With the purity of Russia at stake, as Ilyin would have it, any offensive measures are justified, both at home and abroad. Snyder also placed Russian cyber aggression and disinformation against the US in the context of the struggle between Europe and Russia. As Snyder explained in his book, the main conflict is between the EU and "Eurasia," Putin's preferred vision for the continent. In the proposed Eurasian model, Russia would dominate a massive geopolitical entity stretching from Vladivostok to Lisbon. As a former KGB officer, Putin allegedly believes the fall of the Soviet Union was a great historical tragedy, a wrong that he can right. The Eurasia strategy is Putin's vehicle for rebuilding the glory and influence of the USSR, if not its actual political structure.

The Eurasian vision and its consequent conflict between Russia and Europe are not actually about the United States. The US is merely the military guarantor of EU security. As such, the US must be induced to get out of the way if the Eurasian project is to succeed. Given the intensity of Russia's interest in displacing the EU with Eurasia, it is not surprising that Russia has embarked on extremely aggressive and destructive campaigns against the US.

Snyder, who is fluent in Russian, investigated the publications of the Izborsk Club, a Russian think tank. The club is highly influential on Russian foreign and domestic policy, basing their views on fascist ideology that stresses Russian supremacy in the world. In 2014, the Izborsk Club advocated for a new cold war against the US, with the anticipation that a cyber campaign would be "filling information with misinformation." The goal was "the destruction of some of the important pillars of Western society."[71]

To achieve this destruction of the West, comprising the EU and US, Russia has actively campaigned for the British exit from the EU, the "Brexit," championed by far right political candidates across Europe, from France to Greece. Russia supports sympathetic populist regimes in Poland, Hungary and elsewhere. Russia also appears to have influenced Donald Trump to advocate against NATO—all in the service of destabilizing opponents of Russian expansionism and the creation of "Eurasia."

What will a Russian victory look like? James Kirchick, author of *The End of Europe: Dictators, Demagogues and the Coming Dark Age*, wrote in *Politico*, "Moscow seeks nothing less than a reversal of the momentous historical processes begun in 1989, when Central and Eastern Europeans peacefully reclaimed their freedom after decades of Russian-imposed tyranny." He added, however, "Unlike during the Cold War, Russia seeks not the military and political domination of Europe through the advance of the Red Army and spread of communist ideology, but rather a resetting of the Continent's security order."[72]

Kirchick cited Russian Foreign Minister Sergei Lavrov's comments about the "post-Western world order" in which "Russia's neighbors will have to accept limited sovereignty within a Russian sphere of influence." According to Kirchick, given that the EU and NATO are obstacles to the reassertion of Russian hegemony, "Moscow's long-term strategy is to undermine and ultimately break these institutions from within, thereby neutralizing the concert of nations that has traditionally been necessary to restrain Russian expansion on the Continent."[73]

A countervailing opinion (and there always is such an opinion) holds that Russia's moves are defensive in nature.

George Beebe, who spent 25 years as a Russia analyst at the CIA, argued that Russia feels threatened by what it perceives as American meddling in its sphere of influence. In his book *The Russia Trap*, Beebe makes a strong case for viewing Russia's apparent cyber aggression against the US as retaliatory measures for America's alleged manipulation of Ukrainian elections, taking sides against Russia in Georgia, Transnistria and elsewhere.

At a practical level, it doesn't matter. Whether Russia is deliberately trying to harm the US and West out of some grand imperial design or is merely acting in self-defense, the end results are the same. We are being affected by their cyberattacks and meddling in our political system.

The only way Russia will succeed in dominating Europe and realizing its Eurasian vision is to neutralize the United States. Right or wrong, the US forms the military backbone of NATO. To win, the US must either withdraw from NATO, cease being interested in protecting Europe from Russian aggression, or both.

How bad could this get? No one knows, but comments made by Russian officials are revealing. Timothy Snyder alluded to Russia's efforts to destroy the USA to triumph in its invasion of Ukraine, citing a 2014 comment made by Sergei Glazyev, a Putin advisor: "Glazyev wrote that the 'American elite' had to be 'terminated' for the war in Ukraine to be won."[74]

To achieve this goal, Russia has embarked on a large-scale, multi-faceted and highly ambitious program to make the US quit Europe. As Sciutto noted in *The Shadow War*, the Russian conflict includes many edgy cat-and-mouse games between US and Russian naval and air forces as well as numerous attempts, some successful,[75] that aim to steal national security secrets from the US.

It is possible that Russia will be able to get the US to back away from its security commitments to Europe without causing any lasting political damage to the republic. We may just conclude, based on our own internal political processes, that protecting Europe is no longer a priority for America. For instance, as Elbridge Colby and David Ochmanek, both former deputy assistant secretaries of defense, wrote in *Foreign Policy*, "The United States is trying to defend allies and partners in those other great powers' [Russia's and China's] front yards. The United States' interests in doing so are important, but still partial—and China and Russia's are likely to be considerably deeper. China may well care more about Taiwan, which it considers a renegade province, or Russia the Baltic States, which directly neighbor St. Petersburg, than the United States does about them. This is only natural, but it means that the 'balance of resolve'—which side cares more about the issue—may well favor the other side."[76]

It doesn't seem Russia is willing to wait and see how this turns out, however. Rather, it apparently wants to accelerate the "balance of resolve" either by wearing us out or changing our government altogether. Another option is to make the US so internally preoccupied that it will not have the resources or political will to assert itself in Europe.

From this frame of reference, Russian cyberattacks and disinformation attacks on the US start to make more sense. Though the confrontation between the US and Russia may play out as a long, slow fading out of American influence in Europe, a more rapid decline is probably what the Russians want. Their current strategy and tactics certainly seem to be pushing toward the US suffering a crippling blow of some kind—a downfall that will result in a permanent change in America's geopolitical stance.

Such a downfall might involve a profound change in the way the United States governs itself. If this means making the United States destroy itself in a civil war, so much the better, as far as they are concerned. A USA that will end up killing itself will have zero resolve to protect Ukraine and the whole of Europe.

Timothy Snyder highlighted the reality of this project, writing, "It should be possible, as a Russian military planning document of 2013 proposed, to mobilize 'the protest potential of the population' against its own interests, or as the Izborsk Club specified in 2014, to generate in the United States a 'destructive paranoid reflection.'"[77] The United States in the last three years could be said to be in a state of "destructive paranoid reflection" as Russian propaganda fills our airwaves and political discourse.

For longtime Russia watchers, these moves are anything but new. Such "Active Measures" were a hallmark of Russian/Soviet strategy for decades. In 1998, for example, a retired KBG Major General, Oleg Kalugin told CNN that active measures were "the heart and soul of Soviet intelligence." Kalugin characterized active measures as "subversion, active measures to weaken the West, to drive wedges in the Western community alliances of all sorts, particularly NATO, to sow discord among allies, to weaken the United States in the eyes of the people of Europe."[78]

What's different now is that Russia can digitally reach right into the United States with active measures, at light speed. It can instantly inject disinformation directly into our news media and social media. It can steal information and disrupt American politics at will. Apparently, it can even influence the actions of the American president, who, in just one of many examples of this behavior, parroted Russian propaganda about Crimea by stating, "The people of Crimea, from what I've heard, would rather be with Russia than where they were."[79]

If Russia can keep American presidents thinking and talking this way, it will be able to dominate Eurasia from Vladivostok to Lisbon without doing too much else to us. However, given that the US is still a constitutional republic, Russia will likely have to undertake many more active measures to get what it wants in upcoming election and governance cycles. We can look forward to more disruption at their hands.

China: Understanding the CCP's Strategic Goals

While Trump may give Vladimir Putin the benefit of the doubt, he seems to harbor few illusions about China's threat to the United States. He signed an executive order in mid-2019 banning Huawei and other Chinese telecommunications companies from selling equipment in the US. *Politico* characterized the order as "a move aimed at neutralizing Beijing's ability to compromise next-generation wireless networks and US computer systems." The article cited the potential risks arising from "a foreign adversary" that threatens an "undue risk of sabotage" of US communications systems or "catastrophic effects" to US infrastructure.[80]

The Huawei affair, however, is simply the latest in the long, complicated story of America's relationship with China. Today, China represents a clear strategic threat to the US, though many American leaders in multiple spheres of life have trouble seeing the full scope of the problem. To see where this all came from, and where it might be heading, a bit of history is in order.

Chinese culture dates back thousands of years, and for centuries China was the world's most powerful empire. At the start of the 20th century, however, China was in desperate trouble. Dominated by the British and Japanese empires, the country was powerless and poverty-stricken, riven by civil wars. The Japanese invasion of Manchuria in 1931 and the creation of the puppet state of Manchukuo further humiliated China, which was to endure a torturous Japanese occupation that lasted until 1945.

From 1945 to 1949, the Chinese Communists, led by Mao Zedong, fought a civil war against the Kuomintang (KMT), which was headed by Chiang Kai-shek. Chiang's forces lost that war and retreated to Taiwan, which became known as the Republic of China (ROC). Mainland China became a communist country, the People's Republic of China.

China, despite being the most populous country in the world in 1949, was extremely poor and militarily weak, at least in global terms. And, it was more or less closed off to the Western world for decades. Under Mao's leadership, the country lurched between domestic achievements in areas like healthcare and poverty reduction and disasters like the "Great Leap Forward," a failed modernization plan that led to the starvation deaths of millions of people. The Cultural Revolution, a massive political purge that took place in the 1960s, visited further destruction on a people that was still staggering from decades of war and impoverishment.

The Vietnam War created an opportunity for a reset between the US and China, culminating with President Richard Nixon's visit to China in 1972. This began a slow thawing of relations and a process of modernization. Still, at the time of Mao's death in 1976, China's GDP was $154 billion, or about $321 per capita—China was one of the poorest countries on Earth. When Deng pushed for more modernization after witnessing the defeat of Iraq in 1991, China's GDP was $413 billion, which translated into a per capita GDP of $359 for the average Chinese person.

This history and GDP figures make it easy to see why American policy-makers were slow to recognize changes in China that would enable it to threaten the US in global geopolitical terms. A 50-year-old Washington insider, for example, would have spent his or her college years and early career in a world where China was an economic and military non-entity on the world stage.

Things started to change rapidly in the last two decades, however. Between 2000 and 2018, Chinese GDP leapt from $1.2 trillion to $13.6 trillion, an 11X increase in just 18 years. China is now the second largest economy in the world, poised to overtake the United States by 2030 if its current rate of growth can be sustained.[81]

This explosive growth has occurred because the Chinese Communist Party (CCP), while retaining an iron grip on the political landscape, has ceased to be communist in anything but name. Rather, the CCP is now overseeing a brutal capitalist system that exploits tens of millions of laborers to ensure unbeatable competitiveness for Chinese products in world markets.

The last eight years have seen a significant shift in Chinese policy, following the ascent to power of Party General Secretary Xi Jinping in 2012. As the Hoover Institution's 2018 report, "Chinese Influence & American Interests" notes, "For three and a half decades following the end of the Maoist era, China adhered to Deng Xiaoping's policies of 'reform and opening to the outside world' and 'peaceful development.'"

Now, under Xi, "China has significantly expanded the more assertive set of policies initiated by his predecessor Hu Jintao. These policies not only seek to redefine China's place in the world as a global player, but they also have put forward the notion of a 'China option' (中国方案) that is claimed to be a more efficient developmental model than liberal democracy."[82]

The per capita income in China is up to $7,785,[83] so things have improved up to a point, but unfathomable inequality persists. A small clique of ultra-rich party members reaps the profits of this enterprise. China's 338 billionaires control $6.5 trillion in wealth.[84] Those not so blessed toil in sweatshop conditions, with facilities like the iPhone factory needing to install special nets to prevent workers from killing themselves.[85] This inequality is notable partly because it emphasizes the motivations China's rulers have for staying in power. Getting rich justifies the pain they cause the broader population.

Economic power has enabled China to assert itself geopolitically in ways that would have been difficult to imagine at the start of this century. For perspective, consider that an Annapolis cadet graduating in 2000 would have probably been taught to see China as a potential threat, but not much to worry about. In the 1990s, the Chinese Navy consisted of fewer than 60 frigates and destroyers, accompanied by 80 Soviet-era diesel attack submarines.[86]

When that cadet reached the rank of Admiral in 2019, China had become America's biggest military threat. Its navy comprises over 300 ships, including an aircraft carrier and nuclear submarines.[87] Its air force has been radically modernized, thanks in part to their theft of the F-35 fighter's digital plans in a hack on the Pentagon and Lockheed-Martin.[88]

What is China's plan now, given all this modernization and arms-building? Like Russia, its strategy is at least partly about righting perceived historical wrongs. Robert D. Blackwill, of the US Foreign Policy at the Council on Foreign Relations (CFR) and a former Deputy Assistant to the President and Deputy National Security Advisor for President George W. Bush, wrote in *The National Interest* that "China's primary strategic goal in contemporary times has been the accumulation of 'comprehensive national power.' This pursuit of power in all its dimensions—economic, military, technological and diplomatic—is driven by the conviction that China, a great civilization undone by the hostility of others, could never attain its destiny unless it amassed the power necessary to ward off the hostility of those opposed to this quest."[89]

For Blackwill, "The aims of Beijing's grand strategy both implicate and transcend the United States' and China's other Asian rivals, to replace US primacy in Asia writ large."[90] China's grand strategy is affecting many different areas of the world, but Southeast Asia and the Western Pacific region in particular. China has built man-made islands on formerly uninhabited reefs in the Spratly Islands in the South China Sea, which it is now using as forward air bases.[91]

China claimed in 2015 that it was creating the man-made islands to aid in navigation. Chinese President Xi Jinping assured President Obama that the project was not for military purposes.[92] He soon reneged on that promise, because evidently its plan had been military all along—surprising no one. James Sciutto refers to these new island bases as "unsinkable aircraft carriers."

The Spratly Island bases push China's strategic reach farther out into the Pacific region than ever before. They also threaten Vietnam and the Philippines, as well as a great deal of the world's commercial shipping and the activities of the US Navy's Seventh Fleet. James Sciutto recounted his hair-raising experiences flying on US Navy planes on missions to demonstrate that the US has navigational rights over this international airspace, regardless of China's posturing that the airspace is theirs and theirs alone.[93]

The flights are risky, with Chinese aircraft coming up to challenge the US planes. A collision or downing of one of these planes could be the catalyst for a military confrontation neither side appears to want, at least for now. However, the open conflict is undeniable—a small part of China's much larger policy of "Unrestricted War" against the United States.

In his book, *Stealth War*, Robert Spalding, a retired Air Force General, explained, "The Chinese Communist

Party's ultimate goal, which even the most powerful and well-connected [American leaders] are clueless of or complicit in—is to strengthen itself at every turn. The CCP believes that its biggest obstacle, and indeed, its greatest threat, is the United States of America in as much as it remains the global economic and military leader."[94] Spalding served in a variety of high-level national security roles related to China. Fluent in Chinese, he has spent his career studying the growth of China and its increasing threat to the United States.

As Spalding put it, the CCP's goal and biggest challenge is to displace America's position on the global stage. The displacement may not involve actual military might. "Unrestricted War" is not exclusively military in nature. This is a crucial distinction to grasp if one is to make any sense of China's potential ability to cause a downfall of the US.

China is engaged in a big military buildup. However, as Spalding notes, the CCP's official document describing Unrestricted Warfare states that a nation no longer needs a vast army to conquer. Instead, to control the population, resources and government of the United States, China is employing a variety of non-violent but extremely damaging tactics. These include aggressive cyber espionage against American military targets, massive cyber theft campaigns intended to steal virtually all of America's corporate trade secrets and American civilians' personal data and attacks on the US economy. These will be explored in detail in Chapter 5.

One difficulty in parsing China's attack on the US comes from its success at manipulating American institutions and public opinion without resorting to any subterfuge. As Spalding and others have observed, CCP talking points and Chinese strategy are deeply but invisibly enmeshed in Washington think tanks,[95] the leading American universities, major news media outlets and even in the policies of our elected officials. And, China has enticed—corrupted if you will—a sizable chunk of the American financial and corporate realms. China doesn't have to fight us, so much as buy us, but the risks to the republic are no less grave.

How the US is Reacting to the New Wars

If the US is at war, one might be forgiven for not noticing. We've been tuned into Beatles Brunch, so to speak, while two major powers plot our demise. How did this happen? It's hard to determine the precise cause, but the data is pretty startling.

According to Gallup polling, international problems are the 15th most important issue facing the United States. Just one percent of the Americans surveyed by Gallup consider international matters to be America's most important problem.[96] The number one issue, per Gallup, is immigration, with 25% of respondents rating it as America's most important problem.

Foreign affairs apparently aren't interesting to Americans. Politics has a lot to do with what's interesting to TV audiences, and foreign affairs is considered complicated and not well-suited to the pacing and dynamics of television news. Perhaps it's also because so few Americans are directly affected by international events—at least, as far as they know. Less than one half of one percent of the American population serves in the military.[97] There is no military draft. The county has cut, not raised taxes as it fights wars in Iraq and Afghanistan. So, it would seem that America's foreign adventures are consequence-free for Americans.

Of course, they're not. The US economy and millions of American workers are experiencing the "China Shock," the economic impact of the US trade imbalance with China in the wake of China's admission to the WTO in 2000. Expert opinions vary on the impact of the China Shock, but economists generally believe that about 5 million American manufacturing workers lost their jobs due to the effects of low-wage competition from China.[98]

Thus, although international issues may not seem very important to Americans, perhaps they should be. Alternatively, the China Shock is showing up in Americans' concerns about immigration, with leaders blaming immigrants, rather than China, for job losses.

That said, Americans are not totally clued out about international affairs. A 2017 survey by the Chicago Council on Global Affairs found that 53% of Americans think the US should work to limit Russia's international influence versus 43% who want cooperation with Russia. This is the reverse of the survey findings in 2016.[99]

Conclusion

The US is engaged in geopolitical conflicts against Russia and China that are wars without combat or casualties. Russia calls its process "Hybrid War," while China refers to a strategy of "Unrestricted War" against the US. So, even if we don't want to acknowledge it, our enemies definitely consider themselves to be at war with us. These "win without fighting" strategies involve many different layers of aggression. Cyberattacks and digital disinformation are big components of their victory plans, however. The next two chapters will elaborate on the methods and implications of these threats.

CHAPTER 4

Russia's Digital Dezinformatsiya

The United States is being battered by an endless series of Russian cyberattacks. Russian government groups, and non-governmental entities acting on behalf of the Russian government, are raiding US consumer and government databases. Our critical infrastructure has been attacked, with Russian hackers planting system-destroying malware that can be activated on command. Russian disinformation campaigns pit us against one another and distort our elections. Numerous state and local governments, hospitals and even some law enforcement agencies have been shut down by Russian ransomware.

Is there a unified strategy behind these diverse types of attacks? Based on a review of media reporting, the leadership class of the United Sates doesn't seem to grasp what's happening. For example, out of the 500+ headlines in *The Washington Post* between October 2018 and 2019 that dealt with cybersecurity, none of them mentioned a connection between these separate modes of attack. This is not dispositive, but *The Post* offers a pretty good indicator of how policy elites see things.

Not everyone is silent on this topic. Former FBI counterintelligence and cyber division agent Robert Anderson described the 2016 and looming 2020 election interference to CBS News[100] as "A well-choreographed military operation with units that not only were set up specifically to hack in to obtain information, but other units that were used for psychological warfare were weaponizing that. This is not an operation that was just put together haphazardly."

However, even if reporters believe these attacks are connected, one problem they have is attribution. It's effectively impossible, for instance, to say for certain that a ransomware attack on a city government is from the same source as an Advanced Persistent Threat (APT) strike on the power grid. It's also a speculative story, one that requires imagining events that haven't taken place yet.

All of these obfuscating elements aside, it is still possible to get a cohesive overview of the current situation. This chapter is going to connect dots between the four major types of Russian cyberattack affecting the US right now: 1) data breaches; 2) ransomware; 3) disinformation; and 4) cyber-physical attacks. From these four attack vectors, we can infer a fifth dot: overall Russian strategic intent in these attacks.

It's also worth noting that cyberattacks and disinformation are only parts of broader Russian campaigns of foreign influence. The digital attack fronts should be viewed as synergistic elements of a bigger Russian strategic project that involves espionage, "black ops," military moves, diplomacy, economic pressure and more. Authors and experts with more expertise in these areas can tell that story better than me. The bibliography contains some suggested reading for greater context.

Connecting the Dots

Figure 3 tries to visualize the five vectors of Russian cyber interference directed at the United States. The four outside dots represent known attacks. The fifth dot, at the center, signifies attempts at finding commonality:

- **Dot 1—Russian ransomware**—These attacks are paralyzing American city governments and hospitals

- **Dot 2—Russian data breaches**—Russian forces are stealing massive amounts of information about American citizens, government institutions and government employees

- **Dot 3—Russian disinformation campaigns**—These distortions of reality aggravate, and in some cases cause, racial/ethnic/economic conflict in the US—even to the point of fomenting physical violence based on fiction, e.g. American Muslims want Texas to become a Muslim state

- **Dot 4—Russian "cyber physical" attacks**—This technique uses digital technologies to instigate real-world events, e.g. creating long-lasting power blackouts, destroying industrial production or causing disasters like floods and chemical spills

- **Dot 5—Russian strategic intent**—inferred based on evidence and expert opinion

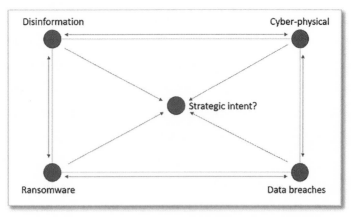

Figure 3 - The "five dots" of Russian cyberattacks on the United States

Getting past Russian deniability

Whodunnit? Russian criminal hackers cracked over 1 billion Yahoo user accounts in 2014. Let's ignore, for a moment, why they would do this. A more important question, for understanding the geopolitical implications of such a data heist, is who was truly responsible? If you believe the Russians, it was simply a criminal matter—hacker gangs trying to make money by selling American account data on the dark web. If you believe the US Department of Justice, the breach was the work of the FSB, Russia's Federal Security Service.[101]

If you want to know for certain, good luck. No one will ever get firm attribution for such incidents. However, many experts have formed a consensus on the culpability of the

Russian government for such criminal acts. An article in *The Deccan Chronicle*, one of India's largest newspapers, quoted a US intelligence official, speaking on condition of anonymity about the Yahoo hack, that "employing criminal hackers helps 'complement Kremlin intentions and provide plausible deniability for the Russian state.'"

The same article then discussed the symbiotic relationship between Moscow's security services and private Russian hackers.[102] It related that US authorities and government cybersecurity specialists have long believed the Kremlin employs criminal hackers for its geostrategic purposes—an arrangement that offers deniability to Moscow and legal immunity for the hackers.

The connections between hacker gangs and the Kremlin are tenuous and unofficial, but they are part of a well-understood environment of government-crime synergy. Organized crime in Russia is tolerated, or even exploited by the government when it suits their purposes. As *The Guardian* reported, "A key characteristic of organized crime in today's Russia is the depth of its interconnectedness with the legitimate economy. Unpicking dirty from clean money in Russia is a hopeless task."[103]

Others have commented on how the Kremlin uses criminal gangs to influence politics in the West.[104] Thus, when Russia denies involvement in hacking, saying it's a criminal matter, this position should be treated with a high degree of suspicion. Jonathan Wrolstad, a cyber threat analyst with FireEye, a firm that follows Russia's intelligence hacking groups, spoke to this issue in *The Hill*, noting, "That symbiotic relationship has been going on for at least 10 years, if not longer."[105]

The article then goes on to cite Tom Kellermann, chief cybersecurity officer at the cyber research firm Trend Micro, who said, "Moscow has long been known to source its technology, world-class hacking talent and even some intelligence information from local cybercrime rings, or 'the Silicon Valley of Eastern Europe.'"

According to Kellerman and others, Russian officials ignore criminality in order to gain an intelligence advantage. As Rep. Adam Schiff (D-Calif.), the top Democrat on the House Intelligence Committee, shared in the article, in Russia, "there are private actors, there are government actors, and there are some that have almost a contractual relationship with the government."

More recently, the connections between the Kremlin and cybercrimes targeting the US have become clearer. In December, 2019, the US Treasury Dept announced sanctions against Evil Corp., a Russian cybercriminal organization. Two of its leading members were indicted[106] in the US for stealing over $170 million in a decade-long spree that took place in more than 40 countries.

American law enforcement accused Evil Corp of being behind the "Bugat," which automates the theft of confidential personal and financial information. A subsequent version of the malware is called "Dridex," which helps install ransomware.[107] Maksim Yakubets, one of the indicted men, had been working for the Russian FSB as late as 2017.108 In other words, his employment by the FSB overlapped his cybercrime activities.

Dot 1–Russian ransomware

Ransomware is a form of cyberattack wherein a hacker implants malware in the target's information systems to encrypt the target's data. This renders the target's information systems useless. The attack also typically takes down email systems, phones, document management software and so forth. The hacker demands a ransom, payable in an untraceable cryptocurrency like Bitcoin, in return for un-encrypting the target's data.

Ransomware can affect anyone. Indeed, many private citizens are attacked in this way, as are corporations, school districts, hospitals and government agencies. As Erich Kron, Security Awareness Advocate at the security firm KnowBe4 put it, "The destructive power of ransomware continues to show how vulnerable organizations are regardless of their size.

It is also a lesson in how long the impact of ransomware can be felt." Kron cited a Kaspersky study that held that 34% of businesses hit with ransomware took a week or more to regain access to their data.[109]

The hackers generally honor the terms of the transaction. They will unencrypt the data they've taken hostage upon payment of the ransom. They may leave malware implants behind, however, which is relevant in municipal cases where the attacker can retain the right to shut the system down all over again at a later date.

In the case of some small cities, their governments have opted to pay the hackers rather than deal with having no municipal services for what could turn into weeks or even months of systemic repair.[110] Cyber insurance policies are also paying ransomware in many cases,[111] though insurance carriers are likely to change their policy terms to avoid giving hackers "easy money" in the future.

Ransomware is affecting government and other public sector organizations. More than 600 government entities, healthcare service providers and school districts, colleges and universities were attacked by ransomware in the first nine months of 2019.[112] According to industry data, these attacks have caused serious disruption to municipal and emergency services, including the diversion of emergency room patients to facilities that are not so affected.

In the summer of 2019, 22 cities in Texas were simultaneously paralyzed by ransomware.[113] Many of these attacks have been attributed to a Russian group known as "Wizard Spider." This gang uses the Ryuk ransomware virus, which has been linked to at least 13 public-sector cyberattacks. Some of these attacks have resulted in six-figure ransom payments by local governments in Georgia, Florida, Indiana and a school district in New York.[114]

In December of 2019, the cities of Pensacola, Florida[115] and New Orleans[116] were hit with ransomware. New Orleans, which was attacked by Ryuk, declared a state of emergency and shut down its computer systems in response. The attack paralyzed the city's courts and the healthcare system for homeless people. City employees switched to Gmail for their work.[117]

The Russian military's GRU Intelligence agency was directly implicated in the 2017 "NotPetya" ransomware attacks that paralyzed over 30,000 laptop and desktop computers and 7,500 servers at Merck, the global pharmaceutical company.[118] The attack affected the company's ability to function for several weeks.

The "Petya" ransomware, a close Russian relative of "NotPetya" disrupted operations at firms like the WPP advertising agency, the DLA Piper law firm, the AP Moller-Maersk shipping line and Pittsburgh's Heritage Valley

Health System, which runs hospitals and care facilities.[119] For perspective on how the US's digital technology is used against it, consider that the Russians created NotPetya and Petya based on the "EternalBlue" Windows exploit developed by the NSA, but later stolen.[120]

The impact of these attacks goes beyond a nuisance and financial losses. A Vanderbilt University research report, cited by the influential KrebsOnSecurity blog, revealed that patient mortality increases at healthcare facilities that have suffered from data breaches or ransomware attacks. According to Krebs, "After data breaches as many as 36 additional deaths per 10,000 heart attacks occurred annually at the hundreds of hospitals examined. The researchers found that for care centers that experienced a breach, it took an additional 2.7 minutes for suspected heart attack patients to receive an electrocardiogram."[121]

What can we make of all this, from the perspective of national security? Some of these attacks are probably just criminal heists. Some are from countries other than Russia. However, it is naïve to assume that sophisticated Russian hacker gangs targeting American public sector entities with ransomware are doing so without the knowledge and approval of the Russian government.

Assuming Russian military and intelligence agencies are behind attacks on hospitals, emergency services and local governments, what would their purpose be? The observable impact of these attacks to date suggests that they want the ability to trigger chaos and panic among ordinary citizens. One might look at the recent spate of attacks as a small-scale trial run of a much larger assault now being prepared for eventual use.

Dot 2—Russian data breaches

Russian hackers, many of whom are presumably linked to the Russian government, have executed cyberattacks and data breaches affecting American business and government agencies. In the process, they have amassed a huge trove of information about American citizens, government institutions and government employees.

For example, a Russian citizen was extradited to the US for hacking American companies and stealing the personal information of 100 million Americans.[122] The man, Andrei Tyurin, 35, was accused of cyberattacks on major US banks and brokerages, including JP Morgan, Fidelity Investments and E*Trade Financial Corp.

Two other Russian nationals were sentenced to prison in the US in 2018 for stealing more than 160 million credit card numbers from credit card processors, banks, retailers and other corporate victims.[123] In 2019, Microsoft revealed that it had detected almost 800 cyberattacks in the previous year that targeted think tanks, NGOs, and other political organizations around the world. According to Microsoft, the majority of the attacks originated in Iran, North Korean, and Russia.[124]

Dot 3–Russian disinformation campaigns

It's beyond question that the Russian government is conducting disinformation campaigns in the US to affect the outcome of US elections. What we see today on Facebook is a familiar pattern. The Soviets called it dezinformatsiya (disinformation). Today, it takes place in high-speed digital form. Should such attacks be perceived as part of a digital attack pattern? Expert opinion is divided on this question. To some extent, disinformation has nothing to do with computers. It's about ideas and influencing popular opinion using the news media and social media. It's also part of a comprehensive intelligence operation.

Indeed, some of the most powerful digital disinformation campaigns are legally (and profitably) streaming into American homes via Russia Today (RT), a cable TV channel funded by the Russian government.[125] However, the superficial legitimacy of RT does not detract from its power to distort American political thinking with Russian propaganda. The channel has been characterized by media analysts as a tool of Russian state propaganda.[126] The Southern Poverty Law Center has criticized RT for pushing the racist "Birther" conspiracy,[127] which effectively launched Donald Trump into national prominence as a politician.

Legitimately placed or not, Russian disinformation campaigns enjoy vast amplification through digital technologies. For disinformation that arrives in the US via social media, though, it's constructive to view the attacks as a form of cyberattack.

The Oxford Internet Institute (OII), for one, sees the issue this way. They published "The Global Disinformation Order" as part of their Computational Propaganda Research Project. The OII report stated, "In 56 countries, we found evidence of formally organized computational propaganda campaigns on Facebook."[128] For context, Facebook disclosed that it deleted 5.4 *billion* fake accounts in 2018.[129]

It may seem like an academic exercise to figure out whether disinformation counts as hacking. The reason it's important to make this determination, however, relates to finding potential solutions. If the problem isn't properly understood, it will be impossible to address. To get some clarity on these issues, I asked a group of industry experts, all of whom have experience working either in or with the US government, to weigh in on this question and others.

Aaron Turner of Highside, an identity security company, took a fairly broad view of hacking. He uses Kevin Mitnick's definition, which includes manipulation of people as a path to hacking machines. As he put it, "Looking at Russian disinformation campaigns, they are designed to manipulate people for them to take certain actions within systems to further Russian foreign policy objectives." To Turner, Russian campaigns do qualify as hacking as they attempt to manipulate others for their own gains.

Richard Henderson is Head of Global Threat Intelligence at Lastline, which uses Artificial Intelligence (AI) to detect cyber threats. He concurred, saying, "Most people, when they think of the word hacking think of someone in a hoody sitting in a dark room actively trying to break into systems. Did Russia 'hack' social media in this way to sow discord in the west? Of course not. But hacking also has another definition—making things do something they weren't designed to do." From this perspective, as Henderson sees it, Russia has definitely hacked major social media to support its intelligence goals. By understanding exactly how these systems operate and how things are quickly disseminated, they have been able to build disinformation campaigns that capitalize on the viral nature of social media.

For Michael J. Covington, Ph.D., VP, Product Strategy at mobile security vendor Wandera, "Any disinformation campaign is absolutely a form of hacking. The Russian attacks aimed at election manipulation are some of the most sophisticated and layered initiatives I've seen, given the diverse systems they're trying to break." Nick Kael, CTO of Ericom Software, a web isolation provider, used Techopedia's broad definition of hacking. This includes attempts to "alter system or security features to accomplish a goal that differs from the original purpose of the system." To him, hacking includes activities such as social media trolling and Twitter bots, and the Russian disinformation campaigns.

Dot 4–Russian "cyber-physical" attacks

Cyber-physical attacks involve hacking computers that control physical systems. The best-known example is the Stuxnet attack on Iranian nuclear facilities in 2010. Though attribution is of course murky, it is believed that a combined American and Israeli cyber force inserted malware into the Supervisory Control and Data Acquisition (SCADA) systems that operated Iran's uranium enrichment centrifuges. The attack simultaneously hacked the centrifuge monitoring system, so operators thought the machines were behaving normally. Instead, by controlling the SCADA units, the Stuxnet virus made the centrifuges spin so fast that they destroyed themselves.

Russian entities, which may or may not be—but most likely are—affiliated with the Russian government, have been engaging in a long series of cyberattacks that have physical outcomes. These attacks include penetration of the American electrical power grid infrastructure and other industrial targets. In June of 2019, the North American Electric Reliability Corporation (NERC), the electric grid regulator issued a warning that a major hacking group with suspected Russian ties was reconnoitering the networks of electrical utilities.[130]

The NERC warning was only the latest in a long series of suspected Russian cyber penetrations of American electrical grid assets.[131] Russian hackers vividly proved their ability to perpetrate a blackout in Ukraine in 2015, when they shut off the computers powering Kiev's electric grid.[132] It is this risk that famed news report Ted Koppel warned about in his bestselling book, *Lights Out: A Cyberattack, A Nation Unprepared, Surviving the Aftermath.*

Can the Russians cause a nationwide blackout? Would it take years to remediate, as Koppel worries—with the potential for civic disruption on an unprecedented scale? According to experts deep in this issue, the massive blackout envisioned by Koppel is unlikely. It would require the successful simultaneous hacking of hundreds of separate corporations. However, these same experts are genuinely worried about the potential for regional blackouts at the hands of Russian or Chinese cyber forces.

The industry is not as well-prepared as it could be,[133] though to their credit, they are working very hard on the issue. Examples of power grid vulnerability appear regularly in the news. For example, security provider Proofpoint reported in 2019, "three small US utilities had been hit with spear phishing attacks a month earlier using the LookBack malware. The malicious emails appeared to impersonate a US-based engineering licensing board, originating from what appeared to be a state sponsored, threat actor-controlled domain."[134] This attack may have been Chinese in origin, a reminder that Russia is not the only enemy we have that wants to shut off our electricity.

SCADA systems tend to be old. They were not designed with cybersecurity in mind. Some of them are "air-gapped," meaning they lack a direct connection to the Internet. Thus, their operators consider them safe from cyberattack. This is less and less true as time goes on. Part of the problem stems from a lack of network awareness. Cybersecurity consultants often do an automated scan of an electrical or industrial facility and find digital assets and Internet connections that the owners had not known about.

The US government is taking cyber physical threats seriously. In December, 2019, the National Infrastructure

Advisory Council (NIAC) presented a report on cyber risks to American critical infrastructure to President Trump. The report, which had been requested by the National Security Council, stated, "Escalating cyber risks to America's critical infrastructures present an existential threat to the continuity of government, economic stability, social order, and national security. US companies find themselves on the front lines of a cyber war they are ill-equipped to win against nation-states intent on disrupting or destroying our critical infrastructure." The report further said, "Bold action is needed to prevent the dire consequences of a catastrophic cyberattack on energy, communication, and financial infrastructures."[135]

The new "Internet of Things" (IoT) also plants sensors and other devices into the SCADA ecosystem and creates unintended and often unknown connections to the Internet. These lightly (or not-at-all) secured devices leave industrial plants badly exposed to cyberattacks that could cause death and destruction.

Getting people to fight each other on the street is arguably another form of cyber-physical attack. For example, researchers at Carnegie Mellon University discovered that 45% of the Twitter messages encouraging people to protest the COVID-19 stay-at-home orders came from software bots, not people.[136] In many cases, these protests involved heavily armed men storming into state capitals and threatening elected officials with death if the lockdowns were not ended.

The researchers could not determine if the Tweets came from American or foreign sources, though the pattern certainly suggests at least some foreign intrigue at work. A few weeks earlier, US intelligence agencies noted that Chinese operatives were spreading fake, inflammatory information about the pandemic in the US via text messages and social media posts.[137] It's not unreasonable to suspect Russian complicity in these events as well.

Russian intelligence entities used Facebook to cause physical unrest in the US during the 2016 election. For example, sixteen thousand Facebook users announced their intention to attend an anti-Trump protest in New York in November 2016. The protest was organized by BlackMattersUS, which was known to be a group linked to Russia. Their goal was to exploit racial tensions in the US. The event was shared by 61,000 people on Facebook. Thousands of people eventually marched in the protest.[138]

Furthermore, as the Senate investigation into 2016 election interference discovered, Russian operatives on Facebook created fake political activism groups that attracted real American members. In of several known incidents, as reported in *The Texas Tribune*, the Russians instigated a confrontation that nearly escalated into a street brawl.[139]

According to the *Tribune*, "Two Russian Facebook pages organized dueling rallies in front of the Islamic Da'wah Center of Houston, according to information released by US Sen. Richard Burr, a North Carolina Republican. Heart of Texas, a Russian-controlled Facebook group that promoted Texas secession, leaned into an image of the state as a land of guns and barbecue and amassed hundreds of thousands of followers. One of their ads on Facebook announced a noon rally on May 21, 2016 to 'Stop Islamification of Texas.' A separate Russian-sponsored group, United Muslims of America, advertised a 'Save Islamic Knowledge' rally for the same place and time."

The *Tribune* then reported, "On that day, protesters organized by the two groups showed up on Travis Street in downtown Houston, a scene that appeared on its face to be a protest and a counter protest. Interactions between the two groups eventually escalated into confrontation and verbal attacks."

The Texas episode was small-scale. No one got hurt. However, its success illustrates the shocking potential of foreign social media manipulation and fake online personas to drive real political violence. After all, the United Muslims event was merely one of many false-but-real political confrontations engineered by Russian hackers on social media during the 2016 election. It would be naïve to conclude that such events would be similarly limited in scope in the future, especially if Russia decided to escalate its digital attacks on the United States.

Dot 5–Inferred strategic intent

Is there a coherent, coordinated strategy behind today's Russian-led data breaches, ransomware attacks, disinformation campaigns and cyber-physical attacks? Expert opinion is divided, but much of the evidence points to at least some strategic commonality. "Russia wants to watch us tear ourselves apart," David Porter, a top agent on the F.B.I.'s Foreign Influence Task Force, told an election security conference in April, 2020. He added, "We see Russia is willing to conduct more brazen and disruptive influence operations because of how it perceives its conflict with the West."[140]

Additionally, *The New York Times* reported in late 2019 on the activities of a previously unknown branch of the Russian military, known as Unit 29155. As the article shared, the unit "has operated for at least a decade, yet Western officials only recently discovered it." *The Times* went on to say that Intelligence officials in four Western countries are unclear how often the unit is mobilized. They warn that it's impossible to know when and where its operatives will strike. However, the existence of this unit underscores Putin's determination to fight the West[141] with hybrid warfare as well as open military confrontation.[142]

The discovery of this military unit is just one aspect of the United States slowly realizing the scale and scope of the cyberwar it's fighting. Professor Roy Godson's testimony to the US Senate includes the estimate that there are between 10,000 and 15,000 Russian operatives engaged in cyberattacks and election interference.[143] The Internet Research Agency (IRA), which influenced the 2016 election, would thus represent less than 1% of Russia's attack capability. If this is true, what are the other 99% doing?

The reality might be even worse than official estimates. According to Adam Darrah, a former US national security professional who now serves as Director of Intelligence at the cybersecurity firm Vigilante, "It would be more than that [10,000-15,000 cyber troops]. The numbers that I know are classified, but I would multiply that a few times. They have a large apparatus that is very good."

I sought additional expert opinions to clarify the matter. Charity Wright, a former military cyber intelligence officer who now serves as a Cyber Threat Intelligence Analyst at Intsights, noted, "They're all tied together. They have these massive military and government sponsored groups and operations that are each specialized in a certain area and in a certain skill and tactics, and so they each have their own objectives. But altogether, they're working toward the greater good of the [Russian] state."

Richard Henderson of Lastline also felt these activities were linked in ways that imply a coordinated effort by state agencies to target the West. As he put it, "There is definitely an overarching strategy by countries such as Russia to wage a low-level cyber war against western governments and corporations. It's part of an overall strategy of asymmetric warfare."

Kristina Libby, an NYU Professor and EVP at the AI-based security firm Hypergiant, felt it would be foolhardy to think they are not connected. As she observed, "Russia is not merely attempting to subvert one area without interest in another." To her, all Russian systems are connected, with ransomware attacks and social media propaganda campaigns working together toward a bigger broader goal.

According to Mike Bittner, Associate Director of Digital Security and Operations at The Media Trust, a digital risk management firm, it was the Soviet leader Nikita Khrushchev who signaled the overall strategy by saying that the USSR would ruin the US from within. Bittner remarked, "As bad actors ramp up their techniques and expand their operations, what might appear to be random incidents are in fact related." Otavio Freire, CTO of SafeGuard Cyber, concurred, sharing that "disinformation and disruptive cyberattacks are all of a piece, though they may not be explicitly coordinated through one body. Indeed, decentralization is the defining characteristic of cyber conflict."

Not all industry experts perceive connections between Russian ransomware, cyberattacks and disinformation campaigns. As Morey J. Haber, CTO and CISO at BeyondTrust, a Privileged Access Management (PAM) company, explained, "I believe they are separate and conducted by two separate threat actors." He felt that social media disinformation is designed to sway public opinion to the goals of the threat actor or cause social discord. He saw no direct financial gains from it, versus ransomware, which is financially motivated. He said, "The threat actors may be opportunistic or targeted, but their end goal is extortion."

According to Andras Toth-Czifra, an Analyst at Flashpoint, which tracks risk intelligence, "It depends on the objectives of the perpetrators of such campaigns." From his perspective, in cases where the goal is to sway public opinion related to a person or a group of individuals, an artificial story or precisely timed social media comment may achieve the desired effect. He did feel it was plausible that social media manipulation campaigns may be used in conjunction with the deployment of other factors, such as DDoS and malware, if the end result justifies the means from the perpetrator's perspective.

Stephanie Douglas, a former FBI agent who now works at Guidepost Solutions, which provides cyber investigations, was not so sure these attacks are coordinated. As she put it, "I would say they're coordinated in that the US is probably their number one target." She was doubtful that there was much actual coordination across different Russian government entities. She did point out, though, that Russian government entities would have to communicate with each other for de-confliction. This is a useful insight. If it's a government program, there has to be some coordination just to make sure organizations aren't getting in each other's way.

Adam Darrah felt the attacks are linked to the extent they help further the Kremlin's goals. However, as he put it, "There's just no bandwidth from the Kremlin, in my opinion, to micromanage every single one of these data breaches of the system penetrations, ransomware attacks and disinformation campaigns."

Based on his experience in the intelligence world, Darrah posited that each separate group in the Russian government is working on its own, with very minimal cooperation because each group is competing for resources domestically. He also offered a twist on this, noting, "They're all trying to bring the biggest 'dead mouse' to their master to see who gets the most money, to be funded further in their efforts against us and against their own people, quite frankly."

Hacking the US to Make Russia Look Good

As Darrah's insight suggests, some of Russia's cyberattacks on the US have nothing to do with the US itself. Rather, these attacks are part of the Russian government's desire to strengthen its own grip on power in Russia. To make its authoritarian state look good, it must make the West look bad. The Russian government also wants to position itself as the victim of attacks from the West. This is at least partly true, though Russia may amplify its victimhood to justify attacks against the US and EU as being retaliatory in nature.

In the view of Yale Professor Timothy Snyder, author of *The Road to Unfreedom*, Putin and his ilk want Russians— and everyone else in their sphere of influence—to see the state as eternal and unchanging. There is no rule of law, no progression to a better future. There is only today. Tomorrow will be no different. Elections and laws are meaningless. Thus, the privileged oligarchs can steal from the people at will, as no arc of history will ensure accountability for wrongdoers. There will never be any change in this state of affairs. This kind of thinking may seem well outside of issues of cyberspace, but cyberwar is one of Russia's main paths to achieving its desired end of history.

Snyder wrote, "In 2013, Russia began to seduce or bully its European neighbors into abandoning their own institutions and histories. If Russia could not become the West, let the

West become Russia. If the flaws in American democracy could be exploited to elect a Russian client, then Putin could prove [to Russians] that the world outside is no better than Russia. Were the European Union or the United States to disintegrate during Putin's lifetime, he could cultivate an illusion of eternity."[144]

"I believe it is part of a broader more sophisticated attack by the Russians to undermine democracies and move sentiment and control towards Russia," said Catherine A. Allen, CEO of Shared Assessments, an organization that helps manage third-party risk. Jeff Williams, the co-founder of Contrast Security, an application security firm, also spoke on the issue by saying, "Of course. All of these Active Measures are tied to fulfilling Vladimir Putin's goal to increase power by disrupting others. Specifically, attacks on technology produce identity information, account access, and kompromat that can be used to create trusted proxies that seem to support their false narratives."

Conclusion

The fog of uncertainty masks Russia's true intentions with its relentless Active Measures campaigns against the US. However, viewing the vectors of attack as a whole, a credible strategic intent emerges. The scale and seriousness of the attacks should make it clear that Russia's goal here is not simply to cause inconveniences for Americans. They are trying to do far more than just disrupt our elections. Whatever they actually do in the end, the depth of the threats we have detected so far highlight the potential for massive political destabilization—even to the point of triggering events that lead to the demise of the Republic—or, as Snyder would have it, the disintegration of the West. Based on Russia's history and its activities in Europe, this objective should not be seen as extreme or paranoid.

CHAPTER 5

China's Unrestricted Plunder

I n December, 2019, the US Navy banned TikTok, a popular Chinese video sharing social media app from government-issued mobile devices.[145] The DoD had issued a similar ban a few weeks later. According to Military.com, the guidance directed all Defense Department employees to "be wary of applications you download, monitor your phones for unusual and unsolicited texts etc., and delete them immediately and uninstall TikTok to circumvent any exposure of personal information."[146] This move, coming after years of devastating Chinese cyberattacks against the Navy, the US government, corporations and civilians, should qualify as a violation of what comedian Jon Stewart once dubbed the "No Sh!t Sherlock Act of 1922."

A week earlier, Senators Chuck Schumer and Tom Cotton, despite being from opposing parties, had jointly called for an investigation into TikTok's security risks.[147] The bipartisan suspicion about Chinese technology hinted at the seriousness of the problem. One might imagine it was the only thing Schumer and Cotton can remotely agree upon at what was a very tense moment in American politics. Still, the US government's response to Chinese cyber aggression was strikingly tepid, especially when viewed against the volume, duration and severity of China's cyber campaigns.

Chinese cyber aggression against the US is not as flamboyant as Russian Active Measures, but it's arguably worse, harming us deeply in ways that we are only beginning to unravel. Part of the problem relates to how much disinformation China can disseminate through open, legal channels like think tanks, financial institutions and politicians who either knowingly or unknowing get money and ideas from the Chinese Community Party (CCP). On the clandestine side, however, the CCP has been impressively industrious. In addition to cyberespionage against the US military, which is a given, there are at least five major vectors of attack:

- Dot 1—Theft of consumer data
- Dot 2—Theft of corporate intellectual property
- Dot 3—Theft of government data
- Dot 4—Theft of military data
- Dot 5—Embedding of malware into Chinese-made technology

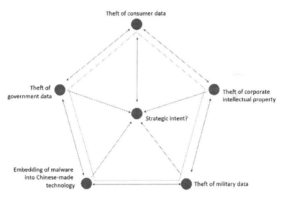

Figure 4 – Connecting the dots in China's five major cyberattack fronts against the US

Figure 4 charts these five dots and asks what their connection is to inferred strategic intent. This exercise is harder than the one described in the previous chapter about Russia. China does not appear to be trying to change election outcomes or openly influence American politics. Yet, the scope and force of their cyberattack fronts suggest its strategic goals are very ambitious.

According to Charity Wright, the former military cybersecurity officer, China's multi-faceted cyberattack campaign "is definitely an organized and coordinated effort as part of China's long-term strategic plan for growth." As she put it, "China has multiple advanced persistent threat (APT) cyber groups that operate somewhat independently from one another." Each group operates like a military unit and has its own area of expertise, specialization, and their own missions/objectives. Some specialize in counterintelligence and espionage. Others develop malware and hacking tools. Some target specific regions, countries, and languages. Others target specific companies, contractors, and government units. According to Wright, these units are all given direction from their military and agency leadership. The orders come from and benefit the state, which is run by the Chinese Communist Party (CCP)/ Communist Party of China (CPC).

Combining the inferred intent of China's multiple cyberattacks with our understanding of Beijing's grand strategy and "Unrestricted Warfare" policy—as well as its slow, insistent military aggression in places like the South China Sea, a bold geopolitical project becomes evident. Wright noted, "Their objective is clear: to take what they consider their rightful place as the 'center country' (中国) and a world economic and political superpower. They use many resources to obtain this objective, including their cyber defensive and offensive operations." China's specific objectives may be elusive, but the broad contours of this project signal serious danger to the American Republic.

Surveillance, Cyberattacks and the Theft of Consumer Data

Anthem Blue Cross, the large health insurance provider, suffered a data breach in 2015 that led to the theft of 78.8 million people's personal information.[148] The company suffered embarrassment and had to pay $115 million to settle lawsuits that arose from the breach.[149] Victims were offered identity theft insurance policies paid for by Anthem.

The US Department of Justice indicted two Chinese nationals for their role in the Anthem breach in 2019. The Anthem breach and its aftermath led *Wall Street Journal* reporters Christopher Porter and Brian Finch to ask a question that was on the minds of many security industry observers: "What Does Beijing Want With Your Medical Records?" As their article noted, "The 2015 Anthem hack raises chilling questions about surveillance within America's borders."[150] China is apparently spying on regular Americans. But why would it do that?

The article offered no viable answer. We don't know why China is stealing tens of millions of private data records on American citizens. What everyone seems to agree on, however, is that whatever the answer is, it can't be good for America. There's no reasonable explanation for why China is stealing huge amounts of data on American citizens. The least harmful excuse I've heard is that China is trying to develop advanced AI-based medical technology so it can dominate the West in this industrial category.

Theft of American consumer data is widespread, according to experts, but as always in these cases, attribution is difficult. The Equifax hack of 2017, which led to the theft of 143 million American credit records, was suspected to be the work of a nation state actor, either Russia or China. One reason that China, or some other state actor, was the prime suspect, was because the hack did not lead to an expected explosion in identity theft cases. Someone stole over a hundred million names, addresses, birth dates and social security numbers but didn't apparently want to use that data to make money.[151] That's suspicious. An American grand jury eventually indicted four members of the Chinese military for the hack in 2020.[152]

Chinese-made consumer technologies have also been caught spying on Americans for years now. For example, as reported in *The New York Times*, at least one brand of Android phones manufactured in China has been shown to send text messages and other data back to China without the user's knowledge or consent.[153] *The Times* reported, "Security contractors recently discovered preinstalled software in some Android phones that monitors where users go, whom they talk to and what they write in text messages."

However, the reason for this activity is not known. The article further stated, "The American authorities say it is not clear whether this represents secretive data mining for advertising purposes or a Chinese government effort to collect intelligence." Some industry experts dismiss the intelligence angle as paranoid. They believe the messaging from devices to China is simply data monitoring for device health and usage for marketing purposes. This may be true, but it's very widespread, in any event. The code from the maker of the phone's software, Shanghai Adups Technology Company, runs on over 700 million phones, cars and other smart devices.

Chinese hackers have also been discovered penetrating telecommunications companies themselves in order to gain access to private communications. As reported in *Dark Reading*, "APT41, a Chinese hacking group known for its prolific state-sponsored espionage campaigns, has begun targeting telecommunications companies with new malware designed to monitor and save SMS traffic from phones belonging to individuals of interest to the government." The article cited research from the cybersecurity firm FireEye Mandiant as the discoverer of these attacks. FireEye would not disclose where the affected phone companies were located, however.[154]

These are just a few of many other suspected Chinese consumer data thefts and acts of surveillance. The news media routinely comes out with fresh stories about surveillance technology embedded in Chinese-made security cameras,[155] smart speakers[156] and smart TVs.[157]

The question, again, is why would they do this? If China wants to defeat the US geopolitically, why would its government need credit and health records? Expert theories on these data thefts cluster around internal Chinese intelligence needs, rather than geopolitics. For instance, some former US government cyber intelligence officials have suggested that China wants to learn if any Chinese citizens are communicating with the Americans. Or, it wants to know, instantly, if any American visiting China is connected with the CIA and so forth—thus, it is building a massive, comprehensive data map of American citizenry.

Theft of Corporate Intellectual Property

China has been conducting cyber raids American corporations for at least the last 10 years. Since 2009, Chinese operatives have been on the hunt for patents, trade secrets, military secrets and information about possible anti-Chinese subversion occurring within American networks like Google's Gmail system.[158] By 2010, at least 34 American companies had been similarly attacked by Chinese actors. According to *The New York Times*, targets included Dow Chemical, Northrop Grumman and Adobe Systems.[159]

Gen. Keith Alexander, as Director of the NSA in 2012, characterized America's loss of industrial information and intellectual property through cyber espionage as the "greatest transfer of wealth in history."[160] He was speaking primarily about China. Alexander claimed that American companies were losing approximately $250 billion per year through intellectual property theft at the time.

Chinese-sponsored data thefts and system intrusions continued unabated through 2014, when then FBI Director James Comey told CBS' *60 Minutes*, "There are two kinds of big companies in the United States. There are those who've been hacked by the Chinese and those who don't know they've been hacked by the Chinese."[161] Comey told CBS that Chinese hackers "were not looking to profit from stolen credit-card details or to find personal information. Instead, they're looking for something similar to trade secrets that can be used in China."

Comey's theory aligns with the opinion of experts like Gen. James Spalding, who wrote in *Stealth War*, "China is utterly fixated on acquiring scientific knowledge."[162] As part of President Xi's "Made in China" plan, China is targeting sectors like information technology, farming machines, electrical equipment, energy-saving cars and aerospace. Spalding noted, "Controlling the design and manufacturing of these sectors is a means to an end. It's the critical first step in subsuming and/or dominating companies across the globe."

As Spalding sees it, by combining industrial dominance with its rapidly growing world-wide shipping and logistics networks, China will "enshrine its ability to ensure geopolitical control over vast swaths of the world."[163] To achieve this ambitious goal, China will do whatever it takes to transfer desired technologies to its companies.

The US government is trying to address this serious attack on the American economy. The US indicted five officers of PLA Unit 61398 for theft of confidential American corporate information.[164] The FBI also warned in 2014 that hackers from China were attacking American companies.[165]

President Obama attempted to address this issue in a meeting with President Xi in 2015, announcing that the two of them "had 'reached a common understanding' to curb cybersecurity and cyberespionage between their countries." As *US News* reported on the meeting, Obama said, "We've agreed that neither the US nor the Chinese government will conduct or knowingly support cyber related theft of intellectual property including trade secrets or other confidential business information for commercial advantage."[166]

Xi's promise was predictably worthless and the Chinese plundering of American corporate secrets continued for years afterward. As the US and China entered into today's current trade war, the attacks escalated. CNN reported in 2019 that "Hackers in China have significantly stepped up attacks on US companies as the two countries have clashed over trade and technology." CNN cited the cybersecurity firm CrowdStrike as seeing "a big resurgence last year in efforts by China-based groups to break into the systems of American businesses for commercial gain—a trend that 'shows no sign of stopping,'" according to CrowdStrike executive Michael Sentonas.[167]

60 Minutes followed two years later with a segment called "The Great Brain Robbery," which cited the Justice Department as saying "the scale of China's corporate espionage is so vast it constitutes a national security emergency, with China targeting virtually every sector of the US economy, and costing American companies hundreds of billions of dollars in losses—and more than two million jobs."

John Carlin, the Assistant Attorney General for National Security with responsibility for counterterrorism, cyberattacks and economic espionage, told Lesley Stahl, "This is a serious threat to our national security. I mean, our economy depends on the ability to innovate. And if there's a dedicated nation state who's using its intelligence apparatus to steal day in and day out what we're trying to develop, that poses a serious threat to our country."[168]

Further evidence of China's unrestricted plundering of American corporate data was revealed by *The Wall Street Journal* in December, 2019. *The Journal* cited an attack known as "Cloud Hopper," which they described as "one of the largest-ever corporate espionage efforts, cyberattackers alleged to be working for China's intelligence services stole volumes of intellectual property, security clearance details and other records from scores of companies over the past several years." According to *The Journal*, "They got access to systems with prospecting secrets for mining company Rio Tinto PLC, and sensitive medical research for electronics and health-care giant Philips NV."[169]

This issue is political as much as it is economic. As American industries suffer, or even shut down due to this massive raising of patented American technology and know-how, regular working Americans are losing their jobs and seeing their standards of living decline. From this, it's a very short distance to resenting the "culprits" who are being blamed for Americans' declining fortunes: Latin American immigrants, "liberal elites" and "globalists," a thinly veiled reference to Jews. Apparently inspired by such messages, a Trump supporter mailed bombs to the homes of George

Soros, Barack Obama and Hillary Clinton in 2018.[170] This attack was ineffectual, but one might imagine a real tragedy, or even a serious civil conflict, arising from this kind of cynical agitation.

Hacking and Data Theft at US Government Agencies

Chinese intelligence operatives have been attacking US government agencies and stealing data in parallel with their raids on American companies. The highest profile incident was the 2015 penetration of the US Office of Personnel Management (OPM), which resulted in the theft of an estimated 22 million confidential personnel records of US government employees. This theft was attributed to China.[171]

The OPM breach included the theft of millions of government employee fingerprints as well as the security clearance application forms of many thousands of Americans working in intelligence and law enforcement.[172] Returning to his role of pointing out painfully obvious truths, James Comey told *The Washington Post* that the OPM breach was "a very big deal from a national security perspective and a counterintelligence perspective. It's a treasure trove of information about everybody who has worked for, tried to work for, or works for the United States government."[173] Chinese access to the security clearance files, which contain sensitive personal and family information, enables China to potentially blackmail US government officials.

While the OPM breach was a high-profile incident, China's attacks on the US government are ongoing. In December, 2019, for instance, the cybersecurity firm Anomali, published a report on phishing attacks against US government agencies. According to Anomali, "The research identified numerous phishing sites designed to steal credentials from victims at 22 government procurement services agencies and several private businesses. Targeted organizations in the United States included the US Department of Energy, US Department of Commerce, US Department of Veterans Affairs, US Department of Transportation, and the US Department of Housing and Urban Affairs."[174]

Hacking the US Military and Stealing Military Secrets

In light of these events, it should seem all the more alarming that the US Navy actually had to issue a memo banning the use of TikTok on government devices in 2019. The risk of such Chinese-made technologies should have been obvious by then. In fact, warnings about Chinese cyberattacks against the US military date back at least to 2007.

While it was not known publicly, the military certainly knew that the digital plans for its top-secret F-35 fighter jet had been stolen by Chinese hackers in 2007. The Edward Snowden revelations brought the matter to the public's attention years later. By then, it was obvious that China's J-20 fighter was a clone of the F-35, though one that is allegedly not as high-performing or advanced.

A congressional advisory group, the US-China Economic and Security Review Commission (USCC), warned in its 2009 annual report that "there has been a marked increase in cyber intrusions originating in China and targeting US government and defense-related computer systems." As *Information Week* reported at the time, this body had previously labeled China's espionage efforts as "the single greatest risk to the security of American technologies" in 2007 and 2008.[175] *Information Week* also noted that the cyberattacks on the Department of Defense information systems increased from 43,880 in 2007 to 54,640 in 2008. Yet, responding to these attacks can be problematic because Chinese espionage and cyber espionage activities may be carried out by individuals without obvious government ties.

According to *Information Week*, a Chinese government spokesperson told the BBC that the report's claims about Chinese espionage are "baseless, unwarranted and irresponsible."[176] Denials notwithstanding, the Chinese hacking of the DoD and defense contractors continued. *The Washington Post* reported in 2010 that "there has been a marked increase in cyber intrusions originating in China and targeting US government and defense-related computer systems."[177]

In 2014, Reuters covered an investigation by the Senate Armed Services Committee that revealed hackers associated with the Chinese government had intruded into information systems at American airlines, technology suppliers and other defense contractors involved with the movement of US troops and military hardware.[178]

The US military's embarrassing procession of cyber admissions continued into 2017, when the US Army ordered its troops to stop using Chinese-made drones. Despite the events of the previous 10 years, it did not seem to occur to the Army that digital hardware made in China might be insecure and able to transmit sensitive military data back to China.[179] The Navy issued a similar memorandum in 2017, though it was not disclosed publicly until 2019.[180] Chinese-made drones are still permitted for general government use, unwise as that may be. A proposed piece of legislation is in the US Senate as of May, 2020, seeking to ban their use.[181]

Then, in 2018, a major breach of a US Navy contractor resulted in the theft of, as *The Washington Post* put it, "massive amounts of highly sensitive data related to undersea warfare— including secret plans to develop a supersonic anti-ship missile for use on US submarines by 2020, according to American officials." Chinese hackers were attacking the DoD's supply chain as well as the Pentagon itself.

According to *The Post*, "Taken were 614 gigabytes of material relating to a closely held project known as Sea Dragon, as well as signals and sensor data, submarine radio room information relating to cryptographic systems, and the Navy submarine development unit's electronic warfare library."[182] For China, a country that seeks to expel the US Navy from the Western Pacific and the South China Sea, this hack was nothing short of a motherlode.

Things didn't get better for the DoD in 2019, unfortunately. In 2019, a study cited by *The Wall Street Journal* showed that the Navy and its industry partners were "Under Siege" by Chinese hackers. The article quoted the study as saying, "Hacking threatens US's standing as world's leading military power."[183] Later in the year, *The New York Times* reported on a criminal fraud case where the US Navy had purchased surveillance equipment that it thought was American made, but, as *The Times* wrote, it was "actually manufactured in China, raising concerns that Beijing could have used it for spying."[184]

Cyber Threats Built into Chinese-Made Technology

A large proportion of electronic devices and computer hardware used in America are made, either in whole or in part, in China. These include:

- Our smartphones
- Our telecommunication hardware, which runs our phone companies' networks
- Our security cameras
- A significant portion of the digital infrastructure that powers our factories and transportation
- Our "smart devices," e.g. smart speakers, drones, digital doorbells, digital thermostats and basically the entire Internet of Things (IoT)
- Our "wearables," e.g. fitness trackers and connected watches
- Our computers and peripherals (e.g. USB drives) and most of our servers and network switches, including those that power Amazon's cloud, Facebook, etc.

Chinese manufacturers of these goods must, by law, have a member of the CCP on their boards. Given this rule, there is frequently no separation at all between China's state intelligence services and the makers of computer gear and consumer gadgets.

Why does this matter? Eavesdropping is one risk, but two further forms of hardware-based cyber intrusion pose more sophisticated threats. Firmware, discussed in Chapter 2, is the software that runs the electronic components of a device. Even if the firmware code is written in the US, if the device is made in China, the manufacturer can usually modify the firmware and insert malicious code into it without anyone knowing.

In a 2015 article titled "Are Your Devices Hardwired For Betrayal?" Cooper Quintin of the Electronic Frontier Foundation (EFF) shared that Kaspersky Lab had released a report on firmware-based attacks, which had previously only been demonstrated in lab settings. He warned readers, "This should serve as a wake up call to security professionals and the hardware industry in general: firmware-based attacks are real and their numbers will only increase. If we don't address this issue now, we risk facing disastrous consequences."[185]

The implantation of malicious hardware is the other threat to consider. In 2018, *Bloomberg* broke a story that shook up the enterprise Information Technology (IT) field, suggesting that motherboards for corporate servers, made in China for SuperMicro, contained tiny spy chips, each about the size of a grain of rice.[186] The article claimed the attack by Chinese spies compromised America's technology supply chain and reached almost 30 US companies, including Amazon and Apple.

Everyone mentioned in the Bloomberg story denied the allegations, called their lawyers and proceeded to shut up. The matter seemed to disappear, but the suspicion that Chinese-made hardware could contain hard-to-spot spy chips caused a stir in the IT field. Researchers later replicated the malicious hardware scheme alleged in the Bloomberg report using chips that cost about $200.[187]

What Will Be the Impact of all These Attacks?

Are these five attack fronts connected? The most honest answer is "we don't know." However, it would be hard to imagine that a major power that has expressly committed itself to a plan to dominate the world economically and militarily is launching such a sustained volume of attacks on a random basis. There would have to be some coherent strategy behind it all, one would at least imagine.

As in the case of Russia, attribution is difficult and often impossible. Coordination between attack vectors and known Chinese government entities is unknowable, at least for people outside the most classified areas of the US government—but those in the know aren't talking about it. So, we're left guessing.

Perhaps the years-long series of attacks on the US public, corporations, government and military are just scattered efforts by unrelated segments of Chinese society. Perhaps an endless parade of malware-laden electronic devices made in China are delivered to the US by happenstance. Certainly, people who view Chinese collective economic aggression as malicious risk being labelled as paranoid, defeatist and counterproductive.[188]

However, regardless of the impetus behind China's cyber assault on the US, the negative impact is still the same. The country is losing intellectual property, which will hurt American business and lead to the further degradation of American workers' economic outlooks. The US military is being weakened badly, no matter how confident it tries to sound in the face of relentless Chinese cyberattacks. The US public is losing more and more of its privacy at the hands of Chinese data thieves and Chinese-made electronics.

Conclusion

China threatens the US, and we are letting them do it with impunity, it seems. As in the case of Russia, it can be difficult to parse so many separate, ostensibly unrelated acts and perceive the scale and force of China's attacks on America. It's happening, though, and shows no sign of abating. Chinese cyberattacks and data thefts are causing catastrophic damage to the US government and military as well as American economic competitiveness. American consumers are subject to extensive invasions of their privacy at the hands of Chinese hackers, with the purpose of these aggressive acts unknown but obviously not benign in nature.

The US government cannot seem to formulate an effective, proportionate response, or even accurately acknowledge what's happening. However, China's unrestricted warfare against the US, of which cyber is a major element, threatens the fabric of American life and politics through economic devastation and potential military defeat.

CHAPTER

6

Nothing's New, Except Everything

King Solomon wrote in the Bible's *Book of Ecclesiastes*, "There is nothing new under the sun." Yet, here we are. So many areas of life, from personal to business to geopolitical, are at once familiar but also completely different. Perhaps Ecclesiastes would allow for the transformative nature of technology. In his day, that might have been the invention of iron weapons versus bronze. Today, it's silicon.

Indeed, Russian disinformation and Chinese espionage are not new. Stalin and Mao were both drivers of all manner of destructive foreign intrigue. Nor did they invent these tactics. They were already old when Alexander the Great tried them out on the Persian Empire in the 4th century BC. Racism is not new either. Economic ups and downs are not new. Greed is not new. Industrial competition between nations is not new. Military arrogance is not new. Slanted journalism is not new.

Today's digital technology, however, manifested across every significant sphere of human endeavor, has wrought a new world. The old patterns and dynamics never change, but technology has revolutionized their efficacy. The economy is different from the way it used to be, as is work, news consumption and life in general. Geopolitics has also changed, radically, under digitization.

The Cold War was largely analog, with offensive campaigns that were comparatively limited in scope and impact. While Cold War dynamics may survive today, they are having a quite different effect on American society and politics. McCarthy-type manipulation and racism are traditional staples of American politics, but today's digital news media and social media give them new prominence and reach.

Nothing's new, it seems, except everything. The United States has been economically weakened by digital technology. The same technology renders the population more susceptible to digital sabotage, disruptions of regular life, fast-moving foreign propaganda and toxic tactics spawned by other Americans. The information sharing dynamics of social media greatly amplifies age-old patterns of public lying. Even the US military finds itself in a more vulnerable position than it wants to admit in this new digital age.

Our Digital World Represents Something Entirely New

Phil Quade, CISO of the multi-billion-dollar cybersecurity vendor Fortinet, captured this idea in his book *The Digital Big Bang: The Hard Stuff, the Soft Stuff, and the Future of Cybersecurity*. He likens the dawn of the Internet era to the cosmic "big bang" theory of the universe—arguing that the Internet, and modern computing in general, comprise a categorically different entity from anything that came before it.

The Internet and modern digital devices move at a higher speed, and offer more connectivity, than any invention in human history. Billions of people worldwide are now connected to one another. With digital devices like smartphones we can instantly access virtually every piece of knowledge generated in the entire history of humanity. Corporate information systems span the globe, linking businesses in complex supply chains and massive, intercontinental logistical operations.

It's all wonderful and dangerous. We are in a new age, full of remarkable potential but also carrying serious risks. To understand our current digital condition, we first need to wrap our heads around the transformative nature of technology, for both good and bad.

Our Collective Failure to Understand the Transformative Nature of Technology

An old friend once told me that we, human beings, are nothing more than apes with car keys. No matter how "advanced" we think we are, our basic appetites, emotions and interactions with one another don't change much. This is true, fundamentally, but technology does alter the human experience.

Technological advances tend to feel familiar because they're often just an update of an earlier technology. The car was merely a faster horse and buggy, for instance. But it wasn't just a faster buggy, was it? The car transformed society, enabling people to live and work differently from anyone who preceded them in history.

Computers offered us a similar arc of disorientation. When they first appeared, they were seen as bigger and better adding machines and filing cabinets. PCs were seen as fancy typewriters. The Internet was a more advanced version of the telephone, and so forth.

As we are all starting to see, though, the computer is more than a big adding machine. The Internet is more than a phone. These are technologies that transform society.

Amazon is not just a bigger, faster mail order catalog business. It's a transformative business, creating new opportunities to work and start businesses. The iPhone is not just a new kind of phone. It's also transformative. It's a pocket-sized computer, camera, audio player, email client, instant messenger—and those are just its basic features. An endless variety of apps make it do a huge variety of other tasks like hailing shared rides, helping us lose weight and on and on.

Transformation is double-edged, of course. The iPhone also tracks our locations and can be hacked to record our voices without our consent. It captures large amounts of our personal data and shares it legally, or secretly, with unknown third parties. Amazon has put scores of competitors out of business and contributed to a reshaping of industries like publishing and retail. An estimated 1.3 million Americans have lost retail jobs in the last 10 years, largely due to competition from Amazon and other online stores.[189]

The Internet and digital technology have rendered the geographic borders of the United States permeable in ways that we are only just beginning to comprehend. While a person must have a passport and a plane ticket to enter the US, crossing through government-controlled gateways to reach his or her destination, digital packets on the Internet face no such barriers. A malware-bearing email from Russia or China faces few obstacles in reaching its target. Russian trolls can insert fake, inflammatory stories into American news media unfettered. Chinese spies can manifest themselves inside US defense contractors and government agencies without any need for plane fares or visas.

So, when Chinese hackers can surface themselves in an American government agency like OPM and steal virtually all of the data it contains, this is no longer "just espionage," as the Obama administration might have had it. When Russian hackers can steal the entire contents of an American political party's email system and publish it to the world, for free, it is not simply just another black bag job like those of yesteryear. These attacks represent the digital transformation of geopolitics.

While American leaders have tended to say the right thing about these dangerous new forms of attack, the follow-up has been poor. President Obama took positive steps like updating the Federal Information Security Management Act of 2002 ("FISMA")[190] to accommodate a leadership role for DHS. However, we still end up, four years later, with a Senate report showing that seven major government agencies are years behind in their FISMA compliance and badly exposed to cyberattacks. The lack of urgency in remediating cyber risk exposure in the government is only partly due to bureaucratic sluggishness. It suggests a lack of awareness of the scale of the danger.

Our lack of insight into digital transformation similarly leads us to overestimate our defenses and resiliency. The current US preoccupation with military-to-military power comparisons in critical regions is a sign of this analytical failure. For example, *The National Interest* ran a piece in 2018 titled "Russia vs. America: Which Army Would Win a War?" This is just one of many such articles that appear regularly in policy circles. In this case, the article is about how "Dismounted firepower is where US forces truly shine."[191]

It goes on to say, "Against armored threats, an American platoon can bring three Javelin missiles, a M3 recoilless rifle, and numerous AT4 short-range antitank rockets to bear against enemy armor, engaging enemies at ranges of up to 2,000 yards." This kind of thinking misses the bigger picture. The M3 may be a good gun, but it won't do much on the battlefield if its digital circuitry can be hacked and rendered inoperable. Even worse, the American army won't even be present if Russia has used digital disinformation to persuade us not to fight at all.

American Society is Digital, as Might Be Its Destruction

We also seem, collectively as Americans, not to comprehend just how reliant our entire society and way of life has become on digital technology. Perhaps it's so obvious that everyone just knows it and it doesn't bear repeating. In my experience, though, writing marketing copy for tech companies and talking to a wide range of people about technology, I have generally found that Americans don't have a good grasp of the subject.

It's not an all-or-nothing proposition of course. In certain areas, people are attuned to technology and its implications. For example, research into attitudes about tech in healthcare shows that Americans are aware of the importance of digital technology in their lives as well as of the risks it poses. A 2019 POLITICO/Harvard poll found that 59% of adults surveyed had used a medical portal to schedule doctors' appointments, but more than half were also "very concerned" their Social Security numbers could be hacked. Less than a third worried about hackers stealing their health information or prescription data.[192]

To be fair, it isn't realistic to expect the average American, or even the average American politician, to understand the breadth and depth of the digital world. It would be good if they did, however. Computer systems, networks and connected devices make possible the proper functioning of every major area of American life, including:

- **Our money and credit cards**—the banking system, and all non-paper money in general, is electronic in form. Though well-defended, banks systems, ATM networks and credit card systems can be hacked and disrupted. A bad banking hack could render us all broke long enough to put the society at risk for mass panic and mayhem.

- **Our healthcare**—nearly every aspect of medicine is computerized now. From scheduling a doctor's appointment to getting a lab test, computers make it happen and keep track of the results. We pay for healthcare through health insurance computer systems, which in turn manage our health claims and pay the medical providers. The hospital experience is almost entirely digital today, with computerized equipment like EKGs and MRI machines connecting to Electronic Health Record (EHR) systems.

- **Our food**—crops may grow in soil, but pretty much everything else about America's food supply relies on computers and computer-controlled equipment.

Food gets from the farm to our tables through computer-driven logistics networks, computerized fuel delivery operations, scientific farming based on big data analytics and artificial intelligence, computer-controlled commodities markets and supermarket procurement software. A bad enough cyberattack on this infrastructure could cause widespread food shortages and attendant social disorder.

- **Our paychecks**—payroll software, which is usually integrated with banking systems and government taxation systems, is what enables us to get paid for our work. Attacks on these systems could keep us from getting paid—not a good situation in a country where so many families are living from paycheck to paycheck.

- **Our heat and electricity**—America's electric grid is computer-controlled at every stage of the cycle that runs from power generation to delivery. The nationwide network of natural gas pipelines and heating oil delivery infrastructure is similarly digital in nature. Hackers can disrupt these services, though at least in these "critical infrastructure" categories, the government and industry have focused intensely on making everything more secure.

- **Our news media**—what we consider "television" and "radio," and to some extent print media, are all digital today. From news capture to editing and broadcasting, it's one continuous digital stream. Today's TV "channels" are just fixed video streams, comparable in format to YouTube videos, just delivered through a dedicated piece of hardware to a computer monitor that we still call a "TV Set." As a result, hackers can take over or black out television networks. This happened in France in 2016, when Russian hackers took TV5Monde off the air by inserting malware into the computer programs that ran the broadcasting operation.[193]

- **Our jobs**—many, if not most of the 157 million Americans in the workforce either use a computer for their jobs or indirectly rely on computers to get their work done. Office workers are closely tied to their digital devices, but workers in healthcare, retail, manufacturing and field service, just to name a few categories, use computers to accomplish tasks.

- **Our corporations and factories**—American businesses are entirely computerized. Information systems manage operations, sales and marketing and distribution. They control logistics, finance and accounting as well as human resources, customer support and so on. In modern manufacturing, most critical systems are Computer/Numerically Controlled (CNC). Products are designed on Computer-Aided Design (CAD) systems and then have their digital plans ported to CNC machines for production. (This is how China was able to steal the designs for the F-35. The plane was, in essence, just a mass of data waiting to be stolen and fabricated in China using CNC equipment.)

- **Our transportation**—travel by air, train and bus depends on computerized scheduling and traffic management systems. Planes themselves are extensively digitized, with the recent tragedies due to software failures in the Boeing 737MAX showing just how crucial digital technology is to aviation.

- **Our cars**—today's cars contain millions of lines of software code[194] to manage their engines and other car systems. Cars are increasingly becoming "connected," with built-in software, circuitry and cellular connections making modern cars constantly linked to their manufacturers and fleet managers. Industry research projects that there will be over 60 million connected cars on the road in 2020.[195] Autonomous vehicles, another emerging trend, are of course totally digital. All of these vehicles rely on GPS, a satellite/computer technology, for navigation.

- **Our government(s)**—at every level, American government organizations run off computer systems. Federal, state and local governments use computers to store data about citizens and government programs, budgets and operational management. Emergency services rely on computerized 911 dispatch software to perform their duties.

- **Our buildings**—modern buildings use computers to control their lighting and HVAC systems as well as elevators and security equipment.

Everything just mentioned can be disrupted or corrupted by hackers. The IoT, for example, experienced a wave of disturbing attacks in 2019. These have included takeovers of smart security cameras, smart TV, smart light bulbs, connected printers and even Wi-Fi-connected coffee machines.[196] Smart speakers and the smart home in general, remain vulnerable to malicious actors.

While it's tempting to ask "who cares?" if someone is trying to hack your smart home or steal your identity through a coffee maker, these attacks reveals the depth of vulnerability we have introduced into our daily lives. They hint at the kind of chaos we might experience if all of these devices and other less visible but important information systems are disrupted. (And, to answer the question just asked, the coffee maker is a gateway into a network that might contain your credit card accounts and so forth, so it does matter.)

Try to imagine what life would be like if every American spent three months trying to survive without money, a paycheck or a job. What if there were no healthcare, pharmacies or hospitals? How would people cope without transportation, government services, emergency services, electricity, heat and food? Add to that toxic blend a lack of news information, or perhaps deliberately wrong, disturbing news—stories planted to incite social groups into violence. Think of this as a "digital blackout." How's it looking? Do you feel safe?

Global Commerce is Digital, to America's Detriment

Political experts have been puzzled by the appeal of Donald Trump to blue collar workers for almost five years. They were befuddled by Trump's remarks about "American carnage" and "putting America first." They wondered why unemployed factory workers in Michigan, to pick just one of many unlikely groups of fans, admired this wealthy New Yorker so much.

These experts seem to be missing two important points. One is that Trump, and other politicians like him, understand that emotional truths are more politically persuasive than facts. The other point has to do with the impact of global digitization on American industry. American workers and politicians may not understand how digitization has affected wages, employment and economic security, but they can definitely see its impact.

What was once considered science fiction in manufacturing and service businesses is now a routine reality: remote employees serving American customers from India and the Philippines; Chinese factories digital integrated into American Enterprise Resource Planning (ERP) systems and global freight and logistics platforms.

In 1990, it would have been hard to imagine manufacturing anything in China except maybe cheap, bulk items like low-end electronics or toys—products that could be made with little need for close communication with the factory. Now, American companies can collaborate in advanced, multi-entity manufacturing chains that run across China in real time. They can communicate across continents using video conference and web presentation software.

The impact of these digital capabilities has been a massive reduction in American jobs, an American carnage, if you will. Almost any American job can be done more cheaply in Asia. This has meant a loss of an estimated 5 million manufacturing jobs and the closing of 22% of American factories between 2000 and 2016.[197] Technology is also affecting the job security of higher-paid white collar workers.

There is an ongoing debate about whether these losses were due to automation or Chinese competition. Some economists argue that it is nonsense to claim that Chinese competition has cost Americans jobs. Rather, they say, automation has put Americans out of work. This can be proved by statistics on worker productivity growth and the rise in American manufacturing revenue during this period.

Statistically true or not, the claim that automation is responsible for American industrial unemployment is masking a huge lie. Thousands of American factories have been shuttered in the last 20 years so their products could be made by hand in China. It's a bit disingenuous to tell millions of out-of-work Americans that a robot took their jobs when the products they used to make by hand in the USA are now made by hand in China, by Chinese workers earning a fraction of their former salaries. It strains one's credulity. One can be forgiven for suspecting that advocates of the automation job loss argument are shills for the CCP or greedy American corporations that wish to cover their tracks.

In some ways, the truth no longer matters. Even though one could present economic data that shows how automation has led to job losses and that wages have grown over this or that period of time, a great many voters feel intuitively that these data points are wrong, even if they're correct.

America's potential downfall, or a whole lot of trouble, may arise out of this disconnect between emotional truths and actual facts. The emotional truth in much of America is that the country is falling apart, devastated by foreign competition and a seemingly indifferent leadership class—a group that does not appear to put Americans' needs first. Again, this is an emotionally compelling and politically powerful issue, even if it's not completely true.

People are angry about the effects of the China shock, even if they don't know what it's called, or that the economic decline they're experiencing is at least in part driven by China and digital technology. The anger is there, nonetheless. An NBC/Wall Street Journal poll in late 2019 found that 70 percent of Americans feel angry "because our political system seems to only be working for the insiders with money and power, like those on Wall Street or in Washington." Forty-three percent say that statement describes them "very well."[198]

This is not an abstract problem. Suicides in America are at their highest level since World War II. As *The New York Times* reported in a shocking OpEd by Nicholas Kristof and Sheryl WuDunn, about the deaths young working-class Americans, "one child in seven is living with a parent suffering from substance abuse; a baby is born every 15 minutes after prenatal exposure to opioids." The authors then made a fascinating association to these data points, writing, "America is slipping as a great power."

Kristof and WuDunn also cited Angus Deaton, the Nobel Prize-winning economist, who said, "The meaningfulness of the working-class life seems to have evaporated. The economy just seems to have stopped delivering for these people." Deaton and fellow economist Anne Case originated the term "deaths of despair" to describe this current surge of mortality from alcohol, drugs and suicide.[199]

Fentanyl, the deadly opioid that kills so many Americans, largely comes from China, exported by criminal gangs.[200] The worst impact of the drug has been felt in the American Midwest, once the center of American industry. It appears that China is targeting the American industry with lethal drugs, perhaps intending to further weaken an economy it wishes to destroy. The history of China, with the horrible

"Opium Wars" of the 19th century, puts the matter into context. Foreigners sent drugs to poison China. Now, they seem to be doing it to us. Experts in the field have told me that the Chinese government is not responsible for this criminal enterprise, but likely tolerates it because it serves their geopolitical interests.

The resulting angry, desperate voters don't go very much for statistical presentations by PhDs. They want results, and they'll go for someone who makes promises that sound good, even if they're completely wrong, counterproductive, racist and cruel. They might even vote in someone whom they can never vote out, an authoritarian president-for-life, an American Putin, so to speak, if they believe this leader will put their needs "first."

The Digital Media Meltdown

Digital technology has brought about a transcendent change in the process of reporting the news as well as the news business itself. The shift started in the analog era, with the rise of television. Neal Postman's landmark 1984 book, *Amusing Ourselves to Death: Public Discourse in the Age of Show Business*, is eerily prescient as it discusses TV's contribution to the breakdown of Americans' ability to grasp and constructively debate policy issues. Long after his death,

Postman's conclusions have been repeatedly re-affirmed with each downward step in public comprehension of political discourse and abuses of the media by demagogues.

The media theorist Marshall McLuhan identified a big part of the problem in the 1960s. His famous quote, "The medium is the message," alludes to the fact that we often focus too much on the content of a new medium while ignoring the fundamental changes in perception that arise from the medium itself. As McLuhan put it, cited by Nicholas Carr in *The Shallows*, "'The effects of technology do not occur at the level of opinions or concepts'… rather, they alter 'patterns of perceptions steadily and without any resistance.'"[201]

Carr further noted, "The showman exaggerates to make his point, but the point stands. Media work their magic, or their mischief, on the nervous system itself." Carr's overall point is that digital technology has changed the way our minds work. His perspective is that the extensive use of the web renders most of us less able to concentrate and parse information as clearly as we did before this technology took over our lives.

The advent of digital cable channels, digital satellite broadcasting has led to an acceleration of news gathering and reporting along with increased competition between news channels. These two factors combine in a negative synergy that lowers the quality of the news while creating messaging echo chambers that reinforce prior opinions.

The story itself starts before digitization, with cable. In 1980, the three major broadcast networks, ABC, CBS and NBC, had a combined nightly news rating of 42.3, which translated into an audience of approximately 45 million television homes.[202] In other words, almost half of America's television households got their news from just three channels.

By 2000, when cable news channels like CNN and Fox News had eaten away at the "big three's" audience, their combined rating was 23.5—a drop of nearly 50%. After going down to 18.8 in 2005, the big three share is back up to about 20 today.[203] Fox News, CNN and MSNBC, the three biggest cable news channels, get about 3 million viewers a day.[204] To summarize: within the span of 20 years, the number of American homes getting their news from major television outlets was cut in half. At the same time, the number of networks providing news to this audience doubled, from three to six.

TV news competition has heated up, so news channels have tried to make their material more flashy and entertaining. This has been accompanied by a corporate mandate to make the news profitable, which never was an issue during the era of network monopolies. These two factors make telegenic and controversial figures like Donald Trump good for ratings. As then CBS Chairman Les Moonves said of the 2016 election and Trump in particular, "It may not be good for America, but it's damn good for CBS."[205]

During this same period, the print journalism business fell apart, done in by the Internet's destruction of classified advertising, one of print's main sources of revenue. Many papers closed and print reporters lost their jobs. Local coverage fell and news quality overall is felt to have suffered, as print reporters were usually the ones doing deep investigative work.

The big newspapers are managing to survive in the Internet age. *The New York Times* website, for example gets over 250 million visitors a day,[206] but the influence of established media sources is on the wane. As a result, the traditional synergy between powerful political figures, respected policy advisors and elite media institutions has all but collapsed.

Part of the problem is simply the extensive segmentation of audiences. Whereas when attorney Joseph Nye Welch asked Joe McCarthy, "At long last, Senator, have you no sense of decency?" in 1954's televised Senate Army-McCarthy hearings, the whole country sat up and took notice. There were only three networks then, so the public dressing down of a powerful Washington figure was a signal event. Today, there are comparable episodes on TV on a regular basis, but few people seem to notice or care, much less be influenced in their opinions. Almost every day, it seems, someone gets "owned" or "schooled" on CNN or Fox, but these events come and go without causing much disruption to the status quo.

Another aspect of the media breakdown is the rise of user-generated content. At once wonderful and terrible, this trend involves regular people creating their own news articles, videos and the like and posting them to websites like YouTube, Tumblr and others. Thus, we have viral videos competing with professionally-produced news. For example, a doctored video that made it seem as if Nancy Pelosi were drunk received over 2 million views on the Internet.[207] That's the equivalent of MSNBC's entire primetime audience.[208]

Sixty-eight percent of Americans get their news from social media today, though 57% say they expect news on social media to be inaccurate.[209] It's a tricky problem to unpack. Sites like Facebook are repeating news from other sources, like TV networks and newspapers, so why won't people trust them? The big takeaway is that America now has few, if any authoritative sources of news that can validate facts and help the public build consensus on important issues.

We also have self-made media celebrities, including the President of the United States. While *The Times'* numbers are impressive, consider that Donald Trump tweeted an average of 15.6 times a day in the first half of 2019,[210] reaching 68 million people each time—a daily cumulative audience of over 1 billion, at no cost to him. That's four times the reach of *The New York Times*, without spending a penny. Each Trump tweet reaches about the same number of people as a Super Bowl ad, which carries an airtime price tag of around $5 million.[211]

Digital technology and the Internet thus enable Trump to buy close to $75 million worth of media for free, every day. His 9,000+ tweets so far in this presidency[212] are equivalent to a roughly $45 *billion* media buy. A more conservative measure, comparing Trump tweets to online advertising, has him achieving 612 billion online impressions—equivalent to an online media buy of $1.45 billion (using a "CPM" of $2.40, an industry average.)[213] That's still a hefty amount of free media and a lot of eyeballs. When leading media and political figures wonder why their messages aren't being heard, while the president is, this market distortion offers an answer.

This is what people are really talking about when they discuss "disruptive technologies." A media market that allows billions of dollars in free time for elected officials is no longer functioning as it used to. One could argue that it's a good thing that ordinary people have broken down the stale old media monopolies. At the same time, as we are seeing, the forces that have taken over are not always good for the public interest.

The lower (or nonexistent) cost of mass media access made possible by social media is only one aspect of its disruptive nature. The other has to do with the breakdown of media relations processes. Traditionally, getting a message out through the news media took time, preparation and professional access. When the Bush administration sought to "sell" the Iraq war in 2003, for example, it carefully drafted talking points and went through the process of booking administration spokespeople like National Security Advisor Condoleezza Rice on major news shows.

This was how administrations got their messages out to the public until our current era. This process is still functioning, but it is increasingly being circumvented. Trump can communicate directly with his audience at will. He does not need professionally-produced talking points and appears to disdain anyone writing them for him. In the cases where he does speak from prepared remarks off a teleprompter, he sounds flat and unconvincing, perhaps deliberately communicating "This is not the real me, the Twitter me."

As established, authoritative media organizations have shrunken, alternative channels have arisen to speak to specialized audience segments. On the left, there are sites like Vox and AlterNet. On the right, we have Breitbart as well as conspiracy theory disseminators like InfoWars and outright neo-Nazi sites like 8chan and Gab. The latter site figured into the mindset of Robert Bowers, the anti-Semite who killed 11 people at a Pittsburgh synagogue in 2018. Discussions on the site had led Bowers to believe that the Hebrew Immigrant Aid Society (HIAS) was importing illegal Latino immigrants to perpetrate a "genocide" against white people.[214]

It's impossible to determine for certain how much foreign influence is driving the content on sites like Gab, but it's undoubtedly present. Russian trolls are active on these sites,[215] but the kind of racist paranoia that pushes people like Bowers into homicidal rages is a very American phenomenon. Russian trolls are simply stoking flames that already burn in this country.

A fascinating *Vox* article, written by Sean Illing on the eve of the Trump impeachment trial, offered insights into just how out of control the media had become in the digital age. He predicted that the Trump impeachment trial would change very few minds in America because, as he put it, "No single version of the truth will be accepted." This was because, "We live in a media ecosystem that overwhelms people with information. Some of that information is accurate, some of it is bogus, and much of it is intentionally misleading. The result is a polity that has increasingly given up on finding out the truth."

Illing cited his colleague, Dave Roberts, who called the media environment of 2019 an "epistemic crisis," where the foundation for shared truth has collapsed. One of the authors of this reality was Steve Bannon, Trump's earliest and most successful strategist, who said in 2018, "The Democrats don't matter. The real opposition is the media. And the way to deal with them is to flood the zone with shit."[216] The "zone" is the digital media ecosystem.

No Needles to Move: The New, High Velocity of Disinformation

Mainstream pundits frequently marvel as the inability of major news stories to "move the needle" of public opinion. This analog metaphor, taken from the old days of Vu meters on audio gear, is exquisitely out-of-date and irrelevant. Americans, it seems, have split into two irreconcilable, hardened groups, each with their own separate views and sources of information, or disinformation as the case may be. There simply are no more needles, even if they could be moved.

Anyone who's dreaded going to a family Thanksgiving meal with relatives from "the other side" will be intuitively familiar with this syndrome. Why is it happening now? American politics and governance have always been messy, dishonest and idiotic. However, there were at least some fact-based controls on the process. This is no longer the case.

In this digital era, disinformation is so fast, persuasive and pervasive that is has altered the political landscape. The old adage about a lie making it around the world while the truth

179

puts its shoes on is now essentially true. Studies published in *The Washington Post*[217] and on NBC News[218] have verified that lies spread faster on social media than the truth. Lies travel at light speed on the net while fusty media types do their fact-checking and broadcast out corrections to any audience that was swayed by the original lie and was no longer paying attention.

The results can be violent, as the infamous "Pizzagate" episode revealed in 2016. After WikiLeaks published emails that Russian forces had stolen from the Clinton campaign, a (likely Russian) campaign on sites like 4chan and Reddit began to foment the conspiracy theory that Clinton and her campaign manager were running a child sex ring out of a Washington pizza parlor.[219] These online threads inspired a man to storm into the pizzeria with a gun and shoot at a door he believed led downstairs to a dungeon of child prostitution. No one was hurt, but the incident showed the power of high-speed disinformation to call Americans to arms.

Why do people believe false stories online? A number of theories attempt to answer this important current question. A study from the University of Minnesota, cited by Fox Business News, demonstrated that cognitive bias affects people's ability to determine whether a story is true or false. They said that participants "only correctly assessed whether headlines on social media were true or false 44 percent of the time." The study found "People were also more likely to believe headlines that aligned with their political beliefs."[220]

Still, false stories have been around since the dawn of time. Why are they so effective and truth-resistant today? One contributing factor is the addictive quality of digital devices. We understand intuitively that our smartphones are an addiction. Research by Deloitte in 2018 showed that Americans check their smartphones 52 times a day on average, an increase from 47 times per day the previous year.[221] Before smartphones, did we do anything 52 times a day on average? And, the actual interactions we have with our phones is far worse. According to *Network World*, we touch our phones 2,600 times a day.[222]

The frequency of use is only part of the story. The companies that make apps are deliberately trying to create addictive conditions. Again, this is intuitively understood by many people, but it's interesting to hear what insiders have to say about the practice. Tristan Harris, a former Google engineer, told Anderson Cooper on *60 Minutes* in 2017, "Well, every time I check my phone, I'm playing the slot machine to see, 'What did I get?' This is one way to hijack people's minds and create a habit, to form a habit. What you do is you make it so when someone pulls a lever, sometimes they get a reward, an exciting reward. And it turns out that this design technique can be embedded inside of all these products."[223]

As Harris explained, programmers call this "brain hacking." The effects of brain hacking are then further compounded by extensive, iterative research on user behavior. Device addiction is a deliberate goal of the technology development process. Data-driven iterations of the process makes the "slot machine" increasingly effective and addictive. From there, purpose-driven data analytics tools can hone the effectiveness of the device to reinforce political messaging and attitudes.

Some have suggested that modern digital devices even enable cult-type indoctrination techniques. Commenting on the current mass phenomenon of people believing demonstrably false ideas, cult expert Steven Hassan said, "As long as a person stays in the cult, they are receiving constant reinforcement of the cult identity." He contrasted today with the 1970s, when cult members had to be taken to an isolated place to have their brains conditioned and programmed by the cult. "Now," he said, "the Internet, YouTube, Facebook and other types of media can program a person into a cult identity."[224]

Companies like Cambridge Analytica mined Facebook user data to identify the best candidates for this kind of experience. They allowed for the optimized delivery of pro-Trump messages to voters whose online behavior suggested they were open to Trump as a candidate. The trend is likely to continue, as Facebook announced in early 2020 that it would continue its policy of allowing political ads to run without the need to verify facts. It would continue to allow the "microtargeting" of users for political ads. Online political ad spending is projected to exceed $1 billion in 2020.[225]

A 2019 book, *The Misinformation Age: How False Beliefs Spread*, laid out a scientific model to demonstrate the impact of such processes. It showed how misinformation results in the hardening of opinion in closed communities like social media groups.

The authors, Cailin O'Connor and James Owen Weatherall, both philosophy of science professors at UC Irvine, illustrated the power of misinformation in social media through a mathematical model known as Bala-Goyal. The model can predict how groups of scientists will change their minds about research over time, based on the influence of peers in academic/social groupings.[226] The model reveals the dynamics of influence that can lead groups of scientists to change their minds or harden their views, regardless of convincing countervailing data.

The authors contrasted two examples of the Bala-Goyal model at work: the evolution of scientific consensus on the depletion of the ozone layer in the 1980s, where scientists came around to a previously debunked theory, versus the current stalemate over global warming. In the latter case, the book shows how corporate propagandists have successfully injected accurate but distortive data into the scientific review process.

They talked about the subtle corporate campaigns to push for certainty in science versus consensus, which can make a big difference to suggestable policy-makers and their voters. There is almost no scientific certainty about anything, so when the corporate propagandist demands certainty, he or she can succeed in pushing a vast amount of legitimate scientific consensus off the table. *The Misinformation Age* revisits the sorry history of the tobacco industry's claims of scientific uncertainty regarding connections between smoking and cancer as an example of this practice.

Social media groupings present a light-speed replication of the biased scientific peer groups in the Bala-Goyal model. People can instantly and permanently believe that the Pope endorsed Donald Trump or that Trump's inauguration crowd was the largest in history—despite obvious evidence to the contrary. O'Connor and Weatherall also describe the power of social acceptance in persuading people to accept false information, even if they are indifferent or don't agree with it.

For example, membership in certain social groups in the US might demand that one declare the theory of evolution to be fake. This creates pressure for members of the group to stick with that story, even if they personally doubt it. Such a dynamic is also at work in modern political discourse. It is then amplified and reinforced by the "slot machine" effect, with thousands of digital bias confirmations happening every week on smartphones.

Then, there's the pace and tonnage of basic news, many elements of which may contradict each other—either by accident, by cynical domestic political operatives or through foreign disinformation. As just one example, consider *Slate* legal analyst Dahlia Lithwick's rundown on a day or two of perplexing news stories coming out of the White House in the aftermath of the assassination of the Iranian military commander, General Qassem Soleimani in early 2020. In this excerpt, I have underlined where Lithwick added hyperlinks to her source stories:

> "Any respite that it [the holiday season] brought, was shattered, first with an extralegal assassination by drone of a state official acknowledged to be an enemy, but not someone whose execution had ever been deemed worth the risk. And with it the flood of hasty lies resurfaced—first that the execution was to stave off an imminent attack that wasn't; then that this was somehow connected to 9/11,

> *which* it wasn't; *then that it was intended to escalate, but also to de-escalate, but also to escalate tensions with Iran. Next came the* tweeted threats to cultural sites, *which were* repudiated, *and then the* memo about troop withdrawal from Iraq, *which was* repudiated, *and then the* White House visit with Saudi Arabian delegates that wasn't, *and the impeachment trial that will proceed without the need for a trial. Having been advised for many months that the military and national intelligence apparatus are 'deep state' liars, we are now told to just trust them again."*[227]

Each story, and contradictory story, has its own digital link. She calls them "hasty lies," which could be a good phrase to describe this entire era. The pattern she is seeing is also a classic Putinesque disinformation technique that involves the simultaneous dissemination of multiple contradictory narratives. While the press sorts them out and argues about it, the deed is done and forgotten. Accountability slips away and we're onto the next crisis. As Americans move through such stories at high speed (and there's no other way to do it if you want to keep up), it's not surprising that we can get confused about what's real.

The impact of false stories can linger long after the truth has been revealed. Timothy Snyder described, for example, how Russia helped instigate and then support the 2014 Scottish referendum on succession from the United Kingdom—an early attempt to split up the EU, later manifesting in the Brexit campaign. Russian media falsely claimed that the referendum vote had contained irregularities, and that the final results (not to leave the UK) were not to be trusted. This was proved to be false, but afterwards, about a third of Scottish voters still believed there had been irregularities.[228]

The Rise of the Digital Party

The changes visited on the US by digital technology and social media have led to the rise of a new political party. It appears to comprise elements of the old Republican Party, but the "Digital Party," as it might be thought of, is something different. It's composed of people who distrust information sources outside of their immediate social groupings, online news sites deemed trustworthy by elite right wing media figures and direct communications they receive from Donald Trump, the leader of the Digital Party.

The Digital Party frightens traditional Republicans, but they are loath to cross this group. In fact, it even appears to be dangerous to criticize the leader of the Digital Party. Critics of

Trump routinely receive death threats from Trump supporters. Congresswoman Maxine Waters and Christina Blasey Ford, who accused Trump court nominee Brett Kavanaugh of sexual assault and "never Trumper" Republican Rick Wilson have all been threatened with death for speaking out against Trump.[229] The death threats are digitally-derived and often include "doxxing," or hacking that publicizes the target's home address, phone number, names of family members and so forth—the better to threaten harm.

Threatening death is an increasingly normalized aspect of American politics. On January 20, 2020, when thousands of heavily armed gun rights protesters descended on Richmond, Virginia, Del. Lee Carter, a state representative who sponsored a gun regulation bill went into hiding, fearing for his safety after receiving death threats.[230] Carter had previously been criticized by gun rights groups on social media, many of whom distorted his proposed gun regulations in online forums. An anti-gun group The Coalition to End Gun Violence decided not to appear, its members having been threatened with violence if they showed up.[231] The rally featured armed protests by Internet-formed groups that are actively advocating for civil unrest,[232] referring to it as a "Boogaloo," which is right-wing slang for war against the government.[233]

Michigan governor, Gretchen Whitmer was similarly threatened with death by (Internet-driven) protestors for enforcing a COVID-19 lockdown.[234] She was singled out for death because she defied President Trump's demand that she reopen the state. He had taken to Twitter, encouraging people to "Liberate Michigan" in 2020.

Conclusion

Nothing's new. Everything's new. We live in confusing times. Our basic geopolitical goals, economic functions and political realities are not all that different from they've been in earlier times. Yet, digital technology has changed the fundamental equation of all of these areas of life. The American economy is staggering under digitally-driven Chinese economic aggression. Our government and political processes have been transformed, not for the better, it would seem, but digital distortions of the news media.

CHAPTER

7

The Risk of a "Cyber 1914"

Aransomware attack took a US Coast Guard base offline for over 30 hours in late 2019.[235] The attack disrupted the operation of security cameras, door access control software and monitoring systems—directly diminishing the capacity of the base to function. It was just one of many recent examples of military assets being revealed as vulnerable to cyberattack. We often hear pundits and government figures warn of a looming "Cyber 9/11" or "Cyber Pearl Harbor." These are legitimate expressions of concern that the US will suffer an unexpected, devastating cyberattack. I would suggest an even more troubling scenario: a "Cyber 1914," in which the US and its enemies get caught up in a cyber war that leads to devastating outcomes no one was expecting.

In August of 1914, Britain, France and Russia went to war against Germany and Austria Hungary. This was the start of World War I. The leaders of these nations expected it to be a brief war, one that would ease a few festering tensions on the continent, leading to a new era of power balance and peace. What they got instead was the deadliest war up to that point in human history. By the end of the war in 1918, over twenty million people were dead and the thrones of Russia, Germany and Austria were empty. The map of Europe was redrawn in ways that more or less guaranteed a second, even worse war twenty-on years later.

What went wrong? While it is well beyond the scope of this book to determine the causes of WWI, it is worth pointing out that the political leaders of the day did not understand the impact of new and as-yet-untested weapons. Hide-bound leaders like Germany's Kaiser Wilhelm had tens of thousands of horses in his cavalry, as did the armies of Britain and France. These horses barely lasted the first few weeks of fighting. Instead, the machine gun and new, breach-loaded artillery conspired to create a lethal stalemate in the trenches.

The war saw the introduction of the tank, the submarine and the airplane in combat, along with poison gas. New, non-military technologies like the telegraph and modern railroads enabled mass troop movements that exceeded the planning abilities of military leaders. Digital technology presents comparable advances in military tactics, though they remain untested in any meaningful war scenario so far. The US military is the most powerful fighting force in history, but it may yet turn out to be shockingly unprepared to fight in a modern war that includes cyber offensive strategies.

The old ways of thinking are less and less relevant, while new vulnerabilities are sneaking up on less than well-prepared armed forces. Like many of the topics in this book, entire books have been written on the subject of the American military's cyber weakness, and prowess. My goal here is not to rehash those other works. Rather, it's to highlight how digital technology has changed the basic equation of military power.

Armies have always tried to play tricks on one another, to get secrets and steal plans. Patton's fake Third Army in East Anglia fooled the Germans into thinking the invasion of France would occur at Pas-de-Calais, not Normandy. The Trojan Horse of ancient times is another great example, one that gave a name to a form of malware.

Militaries also spy on each other, though digital technology has upended any traditional notion of what spying is about. Napoleon had a spyglass so he could see Russian armies on the horizon and move his troops to engage them in battle. But, what if Napoleon could have seen where every single Russian soldier was standing, in real time, so he could aim cannons at every one of them and not miss? That's the difference between analog and digital armies.

America's armed forces, computerized as they may be (and that, in itself is a big weakness), are still rooted in an analog age of warfare. The US military thinks—and spends—in terms of ships, tanks and planes as well as in terms of combat readiness and troop training quality. These are valid factors, but the DoD does not appear to be ready to deal with enemies who use cyber stealth to attack from within.

Many instances of such attacks are now publicly known. Russian spies were able to penetrate the DoD's classified networks by leaving malware-infected USB drives in the Pentagon parking lot. Military personnel evidently picked them up and used them on their PCs, enabling Russian entry into the most secret domains of the US military.[236] White hat hackers at DEFCON 2019 demonstrated that they could penetrate the F-15 fighter's data transfer system due to ports left open by defense contractors.[237]

Reacting to this surprising security gap in a front-line defense asset, the Air Force's assistant secretary of acquisition, Will Roper, said the hackers told him they got into the F-15 through "things industry doesn't know is in their supply chain—it's the ports that weren't cut off, the dry functions that weren't cut off." He added, "Our defense companies are assemblers from the supply chains that they don't require the suppliers to tell them what code, what software functionality is running on components because we don't tell industry to do that."[238]

In other words, the Air Force doesn't fully understand the digital technology it's using to power its most advanced weapons. And, as the hackers showed, it's not hard to detect gaps in the weapons systems and exploit them. The Armed Forces Communications and Electronics Association (AFCEA) made a similar, stark observation about the emerging use of the Internet of Things (IoT) in weapons systems in 2019. An article on the AFCEA website stated, "The military's reliance on the effective, affordable and maturing ecosystem that is the IoT is a plus for efficiency but jeopardizes military supply chains."

As they put it, "A single device such as a mobile phone can contain components manufactured and assembled from various locations around the globe, increasing security risks because of malicious subterfuge by nation-states or even unintentional flaws that yield vulnerabilities... Many devices in the hands of military personnel prove valuable to attackers for several reasons. A smartphone's camera, for example, might provide intelligence about the security of a particular military outpost. Other gadgets provide value because of what they might control."[239]

The military's leadership and foreign policy groups are not unaware of these problems and the potential imbalance between traditional and digital war planning. In 2016, for example, a spokesperson for US Army Cyber Command told *Tech Insider* that cyberspace is "an operational domain: Sea, land, air, space, and cyber. It's a place where our presence exists. Cyber is a normal part of military operations and needs to be considered as such."[240] The excellent book *Cyberspace in Peace and War*, by the Naval Academy professor Martin Libicki, offers a thorough analysis of the military's approach to cyberwar and the processes by which it plans for operations in this realm. It's an eye-opening read for anyone who thinks there are easy solutions to cyber vulnerabilities in the US armed forces.

Further to this point, an article in *Foreign Policy* held that "The 2018 National Defense Strategy shows the Defense Department is focused on the threats posed by Russia and especially China to US interests, allies, and established partners such as Taiwan."[241] According to the authors, US forces appear poorly postured to meet these challenges because both Russia and China have developed formidable networks of missiles, radars and electronic warfare systems. They can degrade and even potentially block US forces' ability to operate in the

Western Pacific and Eastern Europe. This weakens both the US and its allies and partners in those regions.

The article's suggestion that China and Russia could degrade and potentially block US forces' ability to operate is worth underscoring. We have a tremendous military, but it won't do much if it can't function. A 2018 report from the U. S. Government Accountability Office, "Weapon Systems Cybersecurity: DOD Just Beginning to Grapple with Scale of Vulnerabilities," highlights the risks we face. This report is striking partly because of the issues it raises that have been to known to the military for at least five years.

In 2013, the DoD's Defense Science Board released a 146-page report titled "Resilient Military Systems and the Advanced Cyber Threat." The report chronicles years of failed "red team" tests, where ethical hackers easily penetrated DoD networks—as far back as 2008. The report issued a stark warning of a scenario where the US, as they put it, finds "itself in a full-scale conflict with a peer adversary, attacks would be expected to include denial of service, data corruption, supply chain corruption, traitorous insiders, kinetic and related non-kinetic attacks at all altitudes from underwater to space." In such an event, the report warned, American guns, missiles, and bombs may not fire, or may be directed against our own troops. Resupply would be unreliable. Military commanders could quickly lose trust in the information and their ability to control US systems and forces. The reported noted, "Once lost, that trust is very difficult to regain."[242]

The military is also waking up to the fact that non-military cyberattacks may have an impact on the military. The Chinese data breach of the Federal Office of Personnel Management, which involved the theft of millions of military personnel records, gives China an edge in intelligence and beyond. As Joel Brenner, a former top US counterintelligence official, told CBS News, "This tells the Chinese the identities of almost everybody who has got a United States security clearance." For Brenner, and other expert observers, it's a devastating development, making it extremely difficult for any of those people to function as an intelligence officer. The database, which Brenner called a "Gold Mine," also tells the Chinese an enormous amount of information about almost everyone with a security clearance. "It helps you approach and recruit spies,"[243] he said.

The OPM breach also gives China a "gold mine" of data it can use to impersonate military personnel. Combined with voice prints, biometrics and personal data gleaned from innumerable Chinese-made consumer devices, the Chinese can synthesize communications that look as if they're coming from American commanders.

Taking a Hard Look at the USS Fitzgerald Collision

In the context of these breaches and vulnerabilities, it's worth examining some recent events that showcase the potential for military cyber defeat. For example, the cyber threats against the US Navy should cause us to take a hard look at naval accidents that have been attributed to command and training problems, but which also bear the signs of cyberattacks. The most serious of this was the 2017 collision between the destroyer USS Fitzgerald and a 30,000-ton cargo ship, the MV ACX Crystal, in the South China Sea. Seven sailors died in the incident.

I've been warned that I am engaging in irresponsible speculation when I suggest that the USS Fitzgerald crash was due, at least in part, to hacking. Such hacking would logically be Chinese in origin, given the location. To be fair to my accusers, there is no direct evidence of hacking in the Fitzgerald case nor any disclosures by the US Navy. However, a review of the crash and its aftermath should at least raise some questions.

Superb reporting by *ProPublica* revealed that the Fitzgerald experienced a slew of electronic failures in the lead-up to the collision. These included the failure of the ship's email system, which necessitated sailors using Gmail to communicate. The ship's radar failed, so that "One could not be made to automatically track nearby ships. To keep the screen updated, a sailor had to punch a button a thousand times an hour."[244] The ship's navigation system, which ran on Windows 2000 software, also experienced functional problems prior to the crash. Software gets less and less secure as it ages. Hackers discover new flaws in old code. Then, after the maker has stopped supporting the code, as is the case with Windows 2000, there are no more security patches. The software is completely raw and exposed to attack.

As the *ProPublica* story makes clear, the crash occurred because the crew was not sufficiently trained on the equipment—a problem that was compounded by a shortage of personnel and other operational issues. Yet, looking at this tragic episode from a cybersecurity perspective, it's hard not to notice the telltale signs of deliberate hacking, or at the very least, degradation of military systems due to malware. Failures and unexpected bugginess in the email, navigation and radar systems certainly look like malware at work.

In discussions with people knowledgeable about maritime cybersecurity, I learned that the "air gaps" between systems on US Navy ships are not as effective as people in command might believe. According to one source, maintenance personnel can insert USB drives into shipboard systems and implant malware in them even if the system is air-gapped from any Internet-connected system on the ship. The Fitzgerald had recently spent eight months in Japan's Yokosuka's repair yards. Workers there had installed a new defensive system.

Another source revealed to me that US Navy ships are notorious for picking up malware while at sea. Satellite Internet connections enable shipboard systems to pick up viruses while the vessel is at sea. When ships come to port, according to this source, the Naval base's firewalls light up as malware tries to hop off the ship and onto land-based Naval networks. So, it's not a big leap to imagine that malware had infected the Fitzgerald's systems.

The use of Windows 2000 software is also a red flag in this case. This is software that hasn't been supported for security in a decade. When I asked people familiar with the incident if they thought Windows 2000 could be to blame for the Fitzgerald's systemic problems, I got several unconvincing denials.

Within a year of the Fitzgerald collision, the Navy expanded its Cyber Warrant Program to train more warrant officers in cybersecurity.[245] The Navy also launched a new *Command Cyber Operational Readiness Inspections* (CCORI) program in 2018 that allowed, for the first time, for the Navy to conduct and direct mission-based and threat-focused inspections of cyber operations.[246]

The Fitness Tracker Fiasco(s)

US Army personnel accidentally revealed the locations of secret bases in Afghanistan because they wore Strava fitness trackers which broadcast a GPS-based user "heat map" while deployed in January 2018.[247] This was not a Chinese hack, but the incident, and what followed, revealed how hard it seems for the US military to acknowledge cyber vulnerabilities and remediate them. One can only imagine that Russia, China and Iran saw the heat map and studied it to learn as much as they could about the way the US military operates.

Questioned about the lax policies regarding soldiers wearing insecure civilian fitness trackers in secret locations, US Army spokesman, Colonel Robert Manning III said, "We take these matters seriously, and we are reviewing the situation to determine if any additional training or guidance is required, and if any additional policy must be developed to ensure the continued safety of Department of Defense personnel at home and abroad…"[248]

A month later, US Director of National Intelligence, Dan Coats, told the Senate Select Committee on Intelligence, "Frankly, the United States is under attack—under attack by entities that are using cyber to penetrate virtually every major action that takes place in the United States. From US businesses, to the federal government, to state and local governments, the United States is threatened by cyberattacks every day."[249]

Coats' statement reflected a consensus among military commanders and their political counterparts that the United States is engaged in a serious, ongoing cyber conflict. In the background, one could have assumed, the military was moving rapidly to bolster its cyber defenses and strengthen its cybersecurity policies for force protection. This turned out not to be the case, at least when viewed in the context of a subsequent embarrassing, worrisome disclosure.

In July, 2018, six months after the Strava incident, it was reported that the Polar Fitness app, which is used on mobile phones, could be hacked to show the movements and identities of American military personnel and intelligence officers around the world. Using the app's developer API, hackers were able to demonstrate how they could identify, by name, the location and historical movements of over 6,000 people, including staff at Guantanamo Bay and other sensitive sites.[250]

We're Hacking Back

Speaking to current and former military cyber officials about the US military's deficiencies and episodes of cyber damage, I invariably hear the rejoinder, "Oh yes, but we're hacking back. We're doing the same things to them." This is no doubt true. After all, the US is home to some of the best, if not the very best, hackers in the world. However, as the most digitized society in the world, we are also the most vulnerable to cyberattack. We need to be careful how much we play with matches in this tinder box.

It's a good guess that China and Russia can cause the US military more difficulties than we can cause to theirs, via cyber techniques. They don't have the far-flung, cyber vulnerable military contractor supply chains that we do. Their societies are less prone to digital destabilization. China and Russia are also erecting national network perimeters to control or potentially cut off external digital traffic.[251] Human rights advocates protest these moves, saying it stifles citizen freedoms in these countries. That may be true, but it also makes the job of America's hackers that much harder. We have no such protection.

Our Low Level of Spending is Revealing

American military spending on cyber defense also reveals the low status the activity holds in the armed forces. The 2018 budget for the US Cyber Command was $647 million.[252] That's less than one-tenth of one percent of the Pentagon's total budget. The military does spend more than the amount allocated to this one unit. The Navy, Marines, Air Force and Army each have units dedicated to various aspects of cyber defense and so forth. However, even if the total cyber spend was 10X the budget of the cyber command, it's still a tiny 1% of the DoD's overall cash flow. The small spending translates into a limited state of readiness. At a minimum, it reflects a low level of concern about cyber dangers facing the military.

What's Actually the Weakest Point in America's Defense?

It may turn out in the end that the cyber vulnerabilities of the US military are irrelevant. Our enemies may be able to get the US military to pull out of potential fights by manipulating the civilian commander in chief. Speaking to experts on this matter, I've heard several times that the weakest spot in America's defense is its political system.

This would be the ultimate "win without fighting" victory—that the US will simply decide not to fight, ceding Europe to Russian control and the Western Pacific and South China Sea (and Taiwan) to China. The current cyberattacks on the American electoral system and apparent Russian influence on US military policy in Syria, for example, suggest that this strategy may be the most effective of all.

Some Reasons for Cautious Optimism

The US military tends to be slow to change, but they do have mechanisms for addressing their shortcomings and a will to improve. These qualities are on display in the 2020 Cyberspace Solarium Commission Report, which was chartered by the 2019 National Defense Authorization Act. The report contains numerous recommendations for strengthening America's strategic readiness for cyberwar. These include improving deterrence and resiliency. The goals would be to deny benefits and increase costs to foreign cyber attackers. It remains to be seen if the report's recommendations will be implemented. However, the depth of institutional self-reflection revealed in the details of the report is encouraging.

Conclusion

The US is fighting multiple cyberwars at this time. Our enemies appear to have the upper hand, exploiting weaknesses in the military's digital infrastructure and supply chain. While easy solutions are elusive, this grave issue is a low priority. The United States remains vulnerable to military defeat, and attendant political disruption as a result of this risk exposure.

CHAPTER

8

Who Will Not Replace Us?

This chapter is the most speculative in this book, the one that takes me furthest afield from my core experience in digital technology and cybersecurity. It's also intellectually risky to speculate about events that have not yet transpired and take up positions that extend beyond my direct knowledge. My concerns about the future of the country I love compel me to take on this challenge, however. And, I'm not predicting the future. Only a fool or a charlatan would do such a thing. Rather, in this chapter, I will try to demonstrate potential outcomes that could arise from the aggregate risks described so far.

America's reliance on digital technology and its vulnerability to cyber and disinformation attacks—coupled with our enemies' apparent intention to destroy our society and government—make it reasonable to contemplate a digitally-driven political crisis in the US. One could even argue that such a crisis is already underway. Go no further than the events of 2017, when neo-Nazis marched by torchlight in the Unite the Right rally in Charlottesville, Virginia screaming "Blood and soil!" as well as "You will not replace us" and "Jews will not replace us." This ugly episode, after all, was fomented by Russian operatives acting remotely over the Internet.

How much should this event concern us? Upsetting as it might have been, with a hate-fueled murder arising out of the agitated atmosphere, this wasn't the first such march in recent years. There have been events like these for decades. It didn't trigger a massive white supremacist coup d'état, much as its participants might have wanted it to. How do we think history will view Charlottesville? Will it be forgotten, or chalked up as an unfortunate, isolated incident?

Or, will it be seen as an inflection point? Perhaps Charlottesville, with its confluence of neo-Nazism and celebration of The Confederacy represented a moment when the US population began an overt split into two political groups: white and non-white. The blatant, angry but proud racism on display at the march, and other comparable events, might be seen as the start of an inexorable process of destroying the constitutional republic of the United States. The backers of the march want white supremacists in control of the US government. However, it would be challenging in the extreme to have a working constitution, as it now functions, co-existing with a white supremacist monopoly on political power.

This event, and many comparable developments, have led to a situation where 67% of Americans fear a civil war is imminent[253] and the President himself is threatening such a war to his millions of Twitter followers. Post-Charlottesville events certainly suggest such a scenario is in the offing. The protests in 37 American cities over the police killing of an unarmed black man in May 2020 have featured numerous incidents of white supremacists trying to use the protests to foment a race war. The provocations are both physical and digital, with Vice, among many outlets reporting accounts of far-right extremists showing up, with guns, to the protests while others "egging on the violence from behind their computers, urging followers to carry out acts of violence against black protesters with the goal of sparking a 'race war.'"[254]

A few months earlier, Federal authorities arrested Patrik Jordan Mathews, a former Canadian soldier who had entered the US to join "The Base," a well-known and allegedly highly organized and professional neo-Nazi militia. Investigators had recorded Mathews telling his fellow Base members that the January, 2020 gun rights rally in Richmond, Virginia would be an opportunity to start a civil war. He was caught on tape saying, "And the thing is you've got tons of guys who... should be radicalized enough to know that all you gotta do is start making things go wrong and if Virginia can spiral out to (expletive) full blown civil war."[255] According to the BBC, the Base is led by an American named Rinaldo Nazzaro, who directs the group online from a location in Russia.[256]

How did we arrive at this moment in American history? As this book occasionally must, it's time to delve into uncomfortable truths. Since 2000, the US has suffered sustained economic damage due to China's emergence as a global economic competitor. Smart people could argue endlessly about what happened, why it happened and what it means, but the basic reality is undeniable. Millions of Americans have become poorer and had their future economic prospects greatly diminished by the rise of China.

The political environment of the US today, which is increasingly driven by racial and ethnic tensions, is either a direct or significant indirect result of this economic shift. Donald Trump has identified the cause of what he calls "American carnage," and it isn't China. He blames illegal immigrants from Latin America for our plight. He also insinuates that most, if not all, non-white, non-Christian Americans are somehow shrinking the economic pie that is the rightful possession of the "real Americans."

This is dangerous territory for the republic. Angry people are having their desires for authoritarian government stimulated by toxic white nationalist rhetoric. Economically fearful white Americans have become vulnerable to the overtures of a racist demagogue who uses unprecedented digital powers of influence to blame non-whites for their misfortunes. He is telling his followers that their blood and soil, so to speak, are being taken away from them by immigrants and the other. Jews, blacks and Latinos, gays and other various "libtards," as the right-wing media calls their opponents, are the "you" who will not replace the "us," the whites.

Russia is keenly aware of this split in American life. It's delighted to exploit our economic crisis to push for greater and more violent American division. Russia instigated the Charlottesville violence. This fits perfectly with the plan to destabilize the US by aggravating racial fissures in our society. Russia's modus operandi is digital in nature. The goal is the destruction of the United States, or at a minimum, a disruption to American life so severe that it successfully blinds us to Russia's plan to dominate Europe.

Russia and China are poised to visit unbelievable digital destruction on the US. As of now, their capabilities have remained in stealth mode, but the outlines of their attack plans are easy enough to spot. They could involve any number of simultaneous and synergistic attacks on America's power supply, food supply, industrial base and government functioning. The potential for chaos and mayhem is vast.

How will Americans react to such digital triggers? Major weather events and civil incidents like the LA Riots of 1992 offer clues. However, these events took place in a politically stable country. What if mass mayhem caused by cyberattacks were compounded by digital propaganda that aggravated racial tensions—and played into the hands of a would-be authoritarian leader just waiting for such a moment to seize power?

The crisis that ends the existence of the United States as a constitutional republic and removes us from the world stage might feature fewer fireworks than the one just contemplated. It could be a slow devolution of the US that leads to categorical, permanent changes in the way the US is governed and American life is lived. It could involve a slow, grinding erosion of Americans' rights, powers and freedoms.

Indeed, such a scenario may be playing out right now. At the start of 2020, the Trump administration decided to assassinate Iran's top military commander, Maj. Gen. Qasem Soleimani. Though the world will likely not shed too many tears for Soleimani, this act was a clear violation of both

American and international law. The US has rules prohibiting the president from unilaterally assassinating a government official of a country with which we are not at war. The killing took place in Iraq, a sovereign nation, without their awareness or permission.

The president has broad discretion today to initiate military actions without Congressional approval, but the Soleimani assassination should have at least included a briefing of Congressional leaders before it took place. No such briefing occurred. In response, House Speaker Nancy Pelosi is pressing for a Congressional vote on the War Powers Act, which would seek to restrict the president from making such unilateral moves.[257]

Trump responded with a tweet about his decision, declaring that the tweet was his official notice to Congress.[258] A digital message on a politician's personal social media account does not remotely meet the standard for a President's official notice to Congress of the intent to commit an act of war. Yet, he then tweeted that Congress should subscribe to his Twitter feed if they wanted to stay abreast of his war plans and updates on his unilateral military actions.[259]

This move will probably draw meaningless and impotent rebukes from established media and political figures. It will be excused as a joke or "trolling" by Trump supporters. However, the "read my tweet" message is only one of many signals that Trump no longer sees himself as the constitutionally-empowered leader of the republic, subject to the rule of law.

This fear was validated even by members of his own party, who complained that the White House had not been forthright about their reasons for the assassination and evasive about their interest in asking for Congressional approval for the move, and any other such moves. Senator Mike Lee of Utah, a Republican, said that White House officials "were unable or unwilling to identify any point" at which they'd come to Congress for authorization for the use of military force."[260]

Other notable figures expressed concern that Soleimani's assassination portended a breakdown of the constitutional government. Writing in *The Atlantic*, Yale Law School professor Oona A. Hathaway stated, "If Congress fails to respond effectively, the constitutional order will be broken beyond repair, and the president will be left with the unmitigated power to take the country to war on his own—anywhere, anytime, for any reason."[261]

James Webb, the former Senator and Secretary of the Navy, asked in an OpEd in *The Washington Post*, "How did it become acceptable to assassinate one of the top military officers of a country with whom we are not formally at war during a public visit to a third country that had no opposition to his presence?" Webb was mostly criticizing Congress for failing in its constitutional obligations, but he also alluded to the digital breakdown of American politics and culture by observing, "our political debates have come to resemble Kardashian-like ego squabbles."[262]

Trump's Attorney General, William Barr, is advancing the authoritarian presidential agenda, urging for more power to be vested in what he calls a "Unitary Executive."[263] The political commentator David Rohde commented on this strategy in a *New Yorker* piece in early 2020. He wrote, "Barr maintains that Article II of the Constitution gives a President control of all executive-branch agencies, without restriction." Discussing the Trump administration's refusal to honor Congressional subpoenas, Rohde cited Donald Ayer, who served as Deputy Attorney General under George H. W. Bush, who said, "They take the position that they don't even have to show up. That's totally outrageous. It's denying the legitimacy of another branch of government in the name of executive supremacy."[264]

Pew Research suggests that Barr's view is gaining in popularity with the President's supporters. A 2019 survey found that 43% of Republicans "say presidents could operate more effectively if they did not have to worry so much about Congress and the courts," an increase of 16 percentage points from the previous year when only 27% of Republicans held that view. Pew further reported that 51% of Republicans and

Republican-leaning independents "now say it would be too risky to give presidents more power, down from 70% last year."[265]

Is It So Hard to Imagine the Digital Downfall of the United States?

Is it so hard to imagine the demise of the American Republic? The US is one of the longest-running governments in the world. Yet, there's no mandate from heaven declaring that there shall be a constitutional republic in this part of North America. A passing review of recent world history reveals frequent changes in government and national borders on every continent except North America and Antarctica.

With the exception of the UK and Canada, all of the major world economies have had at least two governments in the last 100 years. Since the beginning of World War I, France, Germany, Italy and Russia have had three governments each. (One could even argue that Russia has had five governments since 1914: The Czar, the "Provisional" Kerensky government, the Soviet Union, the Yeltsin government and now the Putin government...) The old East Bloc countries have split up and reorganized politically since 1989. Both China and Japan have had two governments in the last century each.

The success of Russian interference in the 2016 election shows how weak we are, in terms of democratic politics. And, to recap a point made earlier, it was a very minor campaign. Intelligence estimates hold that Russia has between 10,000 and 15,000 military and intelligence personnel dedicated to Active Measures.[266] If 200 of them can help swing a presidential election, what do we think their full force could do to us?

We also forget that we did this very thing to Russia and tried to do it to China. The US pushed events that led to the collapse of the Soviet Union. They're running a similar playbook on us now. Why should we imagine it won't work? China, too, with its long memory, is keenly aware of America's support for the Kuomintang from 1946 to 1949 and the emergence of an independent Taiwan. Why would we imagine they won't try to bring us down?

Current Perspectives on Our Precarious Political Stability

The idea that the United States could suffer a political collapse or an outright civil war is being discussed by high-profile Americans and journalists as well as everyday people. The president himself has made such threats, as has his colleague and former advisor, Roger Stone. Carl Bernstein, the celebrated Watergate reporter, voiced the view the US is in a "cold civil war" during an interview with CNN in November, 2019.

He made his remarks in the context of the impending Trump impeachment, saying the impeachment "will be judged politically in the context of that cold civil war." He advised people to "take a step back, particularly journalists" because "we don't know where this is going to go."[267] The conservative radio host and thought leader Dennis Prager wrote in a 2017 blog post, "It is time for our society to acknowledge a sad truth: America is currently fighting its second Civil War."[268] In his view, the conflict was between the right and left, as opposed to between "rights and liberals," whom he felt could argue and compromise. He believed the left was incapable of compromise, thus triggering this non-violent civil war.

Bill Moyers, one of the elder statesmen of American journalism, expressed his fears of a national collapse in late 2019, telling CNN (as reported in *The Huffington Post*) "for the first time 'in my long life'—including the Depression and World War II—he fears for the nation's survival." Moyers told CNN's Brian Stelter, "a society, a democracy, can die of too many lies—and we're getting close to that terminal moment unless we reverse the obsession with lies that are being fed around the country."[269]

Other thought leaders are implying that a political termination is on its way. Speaking about the alleged conspiracy between Trump, his attorney Rudy Giuliani and criminals connected to the Kremlin, the constitutional scholar Laurence H. Tribe remarked, "If this isn't impeachable and removable conduct, we're done as a constitutional republic."[270]

The respected political author Joel Kotkin wondered out loud in late 2019 if America were about to suffer its "Weimar moment," as he put it, "culminating in the collapse of its republican institutions?" He then said, "Our democracy may be far more rooted than that of Germany's first republic, which fell in 1933 to Adolf Hitler, but there are disturbing similarities."[271]

Supreme Court Chief Justice John Roberts issued a warning in late 2019 that implies a risk to the continued survival of the republic. In his year-end report, he wrote that Americans have "come to take democracy for granted, and civic education has fallen by the wayside." He further pointed out, "In our age, when social media can instantly spread rumor and false information on a grand scale, the public's need to understand our government, and the protections it provides, is ever more vital."[272]

Everyday Americans are expressing similar views. For example, when *The New York Times* interviewed attendees at "Trumpstock," a 2019 political event supporting the president, people admitted to stockpiling weapons in anticipation of Trump losing his 2020 re-election bid. "Nothing less than a civil war would happen," a man named Mr. Villalta told The

Times, which reported that his right hand was reaching for a holstered handgun as he spoke. Villalta added, "I don't believe in violence, but I'll do what I got to do."[273]

An *Esquire* article in 2017 similarly covered a Trump rally in Florida where attendees expressed approval of a possible Trump dictatorship. One person at the rally told Esquire reporter Jeb Lund, "I don't care what he [Trump] does. I'm behind him 100 percent. Put it this way: If he became a dictator, and they said, 'We want him in forever,' he's my man. He's in. I'll never vote against him... I love his power... It's the power that does something to me."[274]

Also in late 2019, a Washington State lawmaker named Matt Shea was accused of being part of a plot by a group known as the "Oath Keepers" to overthrow the US government.[275] *New York Magazine* quoted a tweet by the "Oath Keepers" that read, "This is the truth. We ARE on the verge of a HOT civil war. Like in 1859. That's where we are."[276] Such viewpoints are far from rare in this zone of American political consciousness.

Thus, as we see, thoughts of civil war and political collapse are pervading the political media ecosystem. Americans certainly seem more divided than ever. A 2019 *New York Times* study found that over 42% of people in both parties view their opposition as "downright evil."[277] According to CNN, citing Pew research, "Exit polls from the 2018 midterm elections showed that 76% of voters believe our country is becoming more divided.[278] Over 60% of Americans believe both parties have become 'too extreme.'[279] And 87% of Americans say political polarization is threatening our way of life."[280]

The CNN report further stated, "As the political becomes personal, a quarter of committed conservatives and liberals say that they would be unhappy if a member of their immediate family married someone outside their political party, according to Pew.[281] This interpersonal intolerance is metastasizing into something much darker."[282]

What's driving all of this? The causes of the problem run deeper than the current president and his antics. An article in *The Atlantic*, ominously titled "How America Ends," offered a demographic theory.

> "The United States is undergoing a transition perhaps no rich and stable democracy has ever experienced: Its historically dominant group is on its way to becoming a political minority—and its minority groups are asserting their co-equal rights and interests. If there are precedents for such a transition, they lie here in the United States, where white Englishmen initially predominated, and the boundaries of the dominant group have

been under negotiation ever since. Yet those precedents are hardly comforting. Many of these renegotiations sparked political conflict or open violence, and few were as profound as the one now under way."[283]

What's Driving Worries about a Political Collapse or a Civil War?

Is the US just going through a rough patch, and somehow, in a year, or four, we'll get back to normal? The accelerated, inflammatory nature of digital propaganda may just be burning us out. Is it causing us to fear the worst, when in reality, things are not so bad? Digital technology has put many Americans into an altered state of perception.

Professor Timothy Snyder alluded to this syndrome in his book, *On Tyranny*. He described the uncertainty, the blinking incomprehension of people who spend their days shouting at each other about politics online, actually coming out into the light of day to speak with live human beings about their views. I can certainly relate. The online sphere of discussion, if you could call it that, contributes to a feeling that "there's no talking to them anymore," a feeling of hopelessness and rage.

Bad as all of this may be, it probably doesn't portend an end to the republic. Perhaps everyone is just in a bad state of digitally-stimulated overreaction. There are other explanations for the current, apprehensive moment, however.

Perhaps the original American Civil War never ended

One interpretation of America's state of tension posits that it's not a change of circumstances at all. Rather, this moment of anticipated civil war is simply an uptick in a conflict that never ended. This theory, which has many adherents, views the American Civil War of 1861-1865 as a constant state of war. Indeed, some even believe the conflict predated the official start of armed hostilities in 1861.

Alternatively, even if the war itself ended in 1865, the brutal struggle between blacks and whites, between white power and black subjugation in the US simply morphed into new forms of conflict. After the war, the conflict lived on with the rise of the Ku Klux Klan and the imposition of "Jim Crow" racial segregation in the South. After the Civil Rights era and new federal laws ended these practices, the racial struggle metastasized into the "Southern Strategy" of the Republican Party—a political realignment that saw Southern whites abandon the Democratic Party en masse in the late 1960s. For 50 years, the Republican Party trafficked in racial dog whistles, discussing seemingly neutral issues like urban crime rates, drug crime sentencing and "entitlements reform."

Donald Trump dispensed with the dog whistles, openly calling non-white people rapists and "sons of bitches," retweeted white supremacists and hesitated to distance himself from David Duke, the notorious racist and former Klan Grand Wizard.[284] He also referred to Haiti and African nations as "shithole countries."[285]

The journalist and author Rebecca Solnit spoke to this interpretation of current events in a 2018 article in *The Guardian* titled "The American civil war didn't end. And Trump is a Confederate president." She wrote that the US is in the 158th year of its Civil War. "The Confederacy continues its recent resurgence. Its victims include black people, of course, but also immigrants, Jews, Muslims, Latinos, trans people, gay people and women who want to exercise jurisdiction over their bodies."

Citing the Brazilian philosopher of education Paulo Freire, who said "the oppressors are afraid of losing the 'freedom to oppress'", Solnit noted that the new/old Civil war's premise "appears to be that protection of others limits the rights of white men, and those rights should be unlimited."[286] Solnit's is just one of many similar interpretations of the anxious divide overtaking Americans today. Certainly, these themes figured prominently into the Russian disinformation campaigns and election interference tactics that were on display in 2016 and feared for 2020. These historical themes are still highly potent today. Again, recall that the violent 2017 Charlottesville rally was triggered by a decision to move a statue of the beloved Confederate general Robert E. Lee.

Or, the revolution has already happened

An alternative interpretation holds that the future civil war feared by Americans has already occurred. We're already living in a country that has ceased to be a constitutional republic. We just don't completely know it yet. This may be a bit overblown, but there is some merit to the idea.

Timothy Snyder and others have observed that Donald Trump, who appears to have little regard for democratic institutions and republican government, also frequently seems to obey the wishes of Vladimir Putin, a fascist dictator. (This is not a derogatory label. It is how Putin himself sees his role in Russia and he is quite proud of this fact.) In this sense, the US has certainly taken a step away from self-government by the people.

The idea that the American presidency is less and less democratic is not a new one. The "Imperial Presidency" has been on the rise for a generation or more. Republicans complained about Barack Obama, for example, acting like a king in 2015. After Obama unilaterally declared new national monuments, a freshman congressman from Colorado named

Ken Buck said "He is not king. That is not how we do things in the US Actions like this lead the American people to view Mr. Obama's presidency as an imperial presidency." House Natural Resources Committee chair Rob Bishop further complained that Obama had "sidelined the American public and bulldozed transparency by proclaiming three new national monuments through executive fiat."[287]

More recently, the Fox Business commentator Lou Dobbs was criticized for implying that Americans are obligated to serve Donald Trump. Dobbs said on the air that "It is a shame that this country which is benefitting so much from this president's leadership does not understand their obligations to this leader who is making it possible." In a nearly instant online response that is representative of our generally overheated digital media sphere, *Esquire* columnist Jack Holmes wrote, "The nation has 'obligations to The Leader.' Jesus Christ, man. Really saying the quiet part out loud. In a constitutional republic, the people have no obligation to any politician, not even the president. He works for us."[288]

The digital-psychological nexus at work in American conflict today

Digital technology puts the majority of Americans into close daily (or hourly) contact with politically inflammatory, frequently false messages. We are tightly wired into leadership personalities that are good for edge-of-the seat drama, but not politically or personally healthy. This issue arose in response to President Trump's letter to the House of Representatives after his impeachment in December 2019. The letter asserted, among other things, that the impeachment was an unconstitutional abuse of power, when in fact impeachment is written into the Constitution as a power of Congress.

Salon quoted mental health professionals who felt "This letter is a very obvious demonstration of Donald Trump's severe mental compromise." The author alluded to the digital nature of this risk, saying, "I have been following and interpreting Donald Trump's tweets as a public service, since merely reading them 'gaslights' you and reforms your thoughts in unhealthy ways."

The experts cited in the article advised, "Without arming yourself with the right interpretation, you end up playing into the hands of pathology and helping it—even if you do not fully believe it." The article explains how this is a common phenomenon that happens when people are continually exposed to a severely compromised person without appropriate intervention. The healthy person starts taking on the person's symptoms in a phenomenon called "shared psychosis." The article further warned, "It can also happen at national scale, as renowned mental health experts such as Erich Fromm have noted. Shared psychosis at large scale is also called 'mass hysteria.'"[289]

Mass hysteria seems to be a real risk today, with digital technologies like Twitter and the web fueling its power. According to Steve Schmidt, a respected GOP strategist who quit the party out of disgust with Trump:

> "We're seeing somebody [Trump] go to mass rallies, constantly lie, to incite fervor in a 'cult of personality' base. We're seeing allegations of conspiracy, the deep state, hidden nefarious movements that only the Leader can see. We're seeing the scapegoating of minority populations, the vulnerable population. And lastly, the assertion that 'I need to exercise these powers that no president has ever claimed to have'. This is deliberate. This is an assault on objective truth. And once you get people to surrender their sovereignty—when they think 'what is true is what the Leader says is true' or 'what is true is what the Leader believes is true', even though what is actually true is staring you in the face—when that happens, you're no longer living in a democratic republic." [290]

Of course, we could just be reacting to the fake news

We are so inundated with digital news reports—or at least, people like me choose to be inundated—that one can start to wonder where reality ends and digitally-provoked panic has taken over. Maybe all of this just a big scare perpetrated by digital media companies that get us into a frenzy of online ad clicking and high television ratings. We've never had this much news before, with multiple 24-hour new channels and round-the-cloud news websites churning out alarming stories at a fast clip.

What's true? What's false? Research shows we aren't very good at telling the difference today. Fake news gets mixed in with correct news, often by mistake but just as often by deliberate malfeasance. It would be healthy if we could extricate ourselves from this, but that's harder than it seems. Digital technology is addictive. Reading scary news stories is addictive. Getting agitated by fake stories is part of the addiction.

Russia has been doing its part to plant fake stories that bolster the impression that a civil war is coming. *Newsweek* reported in 2019, for example, that Dmitry Rogozin, a former deputy prime minister and head of Russia's Roscosmos State Space Corporation, cancelled a visit to the US out of fear of "rising tensions between Republicans and Democrats were leading to a breakdown of US society." He told the Rossiya-24

TV channel, "I think that America is actually engulfed by its second civil war now."[291]

An Australian news site similarly reported, "Russia is convinced the next US Presidential election will spark a new civil war." They knowingly added, "And they should know. Moscow has been secretly manipulating US political opinions now for years."[292] As a federal indictment of Russian operatives in the US highlighted, Russian trolls were specifically instructed to amplify the second civil war story. NPR reported on this case, noting, "Influence specialists also got directions in real time about how to respond to specific stories. On Aug. 6, 2017, for example, they received a link to a story quoting radio host Michael Savage promising that there would be a 'civil war' if Trump were 'taken down.'"[293]

Then, there's "Yes California," a political action committee promoting the secession of California from the United State. The project is backed by an American Trump supporter who lives in Russia. It is evidently also supported by the Russian government.[294]

Ultimately, maybe it doesn't matter whether news is fake or not, or whether we've overloaded ourselves. The political outcome will be the same: Americans are on edge, primed for extreme reactions, perhaps even violence, at the hands of digital overstimulation, power-hungry politicians and their irresponsible enablers in the digital media.

Possible Scenarios of a Digital Downfall

What would a digital downfall look like? The odds are that such an event will never take place. However, the risks of a political collapse or a major change in government appear to be higher now than perhaps at any time since 1859. In my view, it's worth mapping out a number of possible downfall scenarios in order to understand how they might be averted.

The basic scenario: a slow fade out

The United States is in the midst of an economic and political shift. This is not a paranoid theory, but rather a rational reading of the facts. People are getting poorer, in terms of the value of their real incomes, and more economically insecure over time. This trend is not likely to improve if current conditions prevail. In parallel, the country has become more ideologically polarized, with each side experiencing a digitally-driven hardening of its views.

The downfall could be slow and hard to see. It would be a steady erosion of people's rights and freedoms, a creeping authoritarianism. One could say this is well on its way to happening, given changes in government surveillance practices and the like.

The declining economic prospects of working Americans have already been shown to expose them to racially exploitive political messaging. Given that poor whites in rural areas wield disproportionate political power due to constitutional distortions to democratic representation, this risk could blossom into a more racially repressive, less democratic and free United States.

We might still live in the United States, but it would be less and less of a constitutional republic. With the inherent racial distortions of the electoral process, aggravated by data-driven gerrymandering and mischief and digital distortions of public opinion, white nationalist politicians would represent white people only. The opposition would be repressed by doxxing and violent, digital vigilantism. The president would serve a small, powerful group of white voters, ignoring Congress, the law and the federal government to rule by personal fiat, as delivered via tweets. Non-white Americans would be second-class citizens, with little voice in how the country is governed.

Over time, the country would become weaker economically and withdraw from regions where it threatened Russia and China. The US would cease to be the number one military power in the world, a transition that might further negatively affect the American economy—worsening a cycle of poverty and repressive identity politics.

Digital triggers for political restructuring

Plots to overthrow the republican government of the United States have been undertaken before. In 1933, for example, the "Business Plot" allegedly involved wealthy men recruiting the respected Marine General Smedley Butler to lead a sham veteran's organization that would stage a coup d'état. There's always something like this afoot, so maybe we shouldn't worry. Yet, as we are seeing, digital technology changes the levers of power, the susceptibility of the public to disinformation and even the nature of reality itself.

Could a cyberattack trigger a political revolution in the US? I think the answer, even in the event of a single catastrophic attack, is no. However, under the right scenario, a combination of cyberattacks, disinformation and willing collaborators, digital interference in this country could lead to disastrous, republic-ending events.

Let's build a scenario. This is what cybersecurity and military strategists do. They model potential events and try to predict reactions and outcomes. Indeed, the Department of Homeland Security engaged in this sort of exercise in 2019. They tasked Cybereason, an Israeli cybersecurity firm, with creating digital simulations of foreign interference in the 2020 election. One scenario they modeled resulted in the hypothetical deaths of 32 Americans, with 200 injured—and the cancellation of the election and declaration of martial law.[295]

Intent is a critical factor to model in building a scenario for America's digital downfall. If a foreign cyber actor wanted to disrupt American life, that's different from an actor trying to foment a political revolution in this country. People fret about attacks on the US power grid, for example. An electric blackout in a major city would certainly be disruptive, potentially even leading to looting and civil unrest. It would probably play out like a bad hurricane. If that were the extent of the damage, the republic would survive.

What happens, though, if the attacker devises a multi-pronged attack intended to cause as much politically charged mayhem as possible? For example, let's imagine that Congress passes drastic, veto-proof sanctions on Russia for its war against Ukraine. Putin, fearing the sanctions could affect his ability to maintain power, unleashes an attack to cause the US to stop functioning as a nation for a prolonged period. This would solve his sanctions problem and take the US out of Europe so he could pursue his dream of dominating Eurasia.

Imagine a scenario more elaborate than the "tabletop" exercise conducted by Cybereason, which simply asked different federal agencies what they would do under various hypothetical situations. Here, Russia launches cyberattacks that cause long-lasting (e.g. transformer-exploding) power blackouts in New York, Chicago, Los Angeles and Boston. At the same time, it instigates a digital blackout, a cessation of online services coupled with widespread computer malfunctioning. All commercial digital services, such as the ones that power gas stations, supermarkets, logistical supply chains, banks, ATMs, natural gas pipelines, governments and hospitals are all switched off or paralyzed by ransomware. Connected cars won't start. 911 services are disabled. Police and fire cannot communicate with one another or function at all, their equipment permanently "bricked."

This may sound like a bad sci-fi movie, but the Pentagon is aware of this potential scenario. The Defense Science Board's 2013 "Resilient Military Systems and the Advanced Cyber Threat" report warned, "The impact of a destructive cyberattack on the civilian population would be even greater [than on the military] with no electricity, money, communications, TV, radio, or fuel (electrically pumped). In a short time, food and medicine distribution systems would be ineffective; transportation would fail or become so chaotic as to be useless."

How would this play out? In the DoD's view, "Law enforcement, medical staff, and emergency personnel capabilities could be expected to be barely functional in the short term and dysfunctional over sustained periods." They further warned, "If an attack's effects cause physical damage to control systems, pumps, engines, generators, controllers, etc., the unavailability of parts and manufacturing capacity could mean months to years are required to rebuild and reestablish basic infrastructure operation."[296]

As this unfolds, the Russians also release a massive disinformation campaign. It consists of "deep fake" videos of democratic politicians urging black men to rape white women. Millions of suspected white nationalists receive text messages that say blacks are killing whites in their town—that they should grab their guns and deal with the crisis. In tandem, millions of black people get text messages claiming that white lynch mobs are roaming the streets, and that the police are in on it.

In Texas, white supremacists identified through Big Data analytics receive messages that liberal "gun grabbers," paired with Antifa brigades and armed Muslims, will be stopping cars at random to seize lawfully-owned firearms. In California, citizens will hear that their water supply has been poisoned by Jews, with a suggestion that those responsible be "made to pay for it."

I realize this is a potentially offensive scenario, but as we know, Russia has no problem "going there." It knows all about American racism and white racial fears. It knows what buttons to push to get white people reaching for their ample supplies of guns and ammunition.

Recent stories suggest that such violence is lurking just beneath the surface, ready to explode. In 2019, for example, the right-wing Christian Pastor Rick Wiles warned that "President Donald Trump's supporters will 'hunt down' Democrats and bring 'violence to America'" once the president leaves office.

As *Yahoo News* reported, "On his apocalyptic *TruNews* program, captured by Right Wing Watch, Mr Wiles said the president's impeachment or 'however he leaves' office will inspire 'veterans, cowboys, mountain men' and 'guys that know how to fight' to bring 'violence to America' by hunting down Mr Trump's political enemies." Wiles' co-host, Edward Szall, said "once the blood starts flowing, it's near impossible to stop."[297] TruNews reaches more than 22 million people.[298]

As the hypothetical scenario unfolds over the course of two weeks, during which the majority of American families will run out of money, food, fuel and heat, there is mayhem in the street. Looting ensues. There is no accurate information available to anyone. Rumors drive digitally-accelerated fear and panic. All news channels are either digital blocked or airing fake, highly inflammatory and digitally-localized news. The veterans, cowboys and "guys who know how to fight" are urged by their pastor to take to the streets and kill the Democrats. The National Guard is called in, but most of them don't know they've been summoned because they're offline.

Those who do report for duty, though, have equipment that won't work because it's been hacked.

Cybersecurity teams desperately try to remediate the problems. Not every system has been breached, but the destruction is so interconnected and pervasive, that they cannot get the major systems back online. Plus, the team members themselves fear for their safety. Many choose to stay home and protect their families. Nor can they communicate with one another, as wireless telecommunication networks have been hacked and no calls or texts will go through.

What will a would-be authoritarian leader do in response to such a crisis? Would he calmly urge Americans to stay home and avoid violence? Would he or she summon Congress and devise a workable emergency response?

He and his supporters have been loudly arguing for a "Unitary Executive," i.e. non-constitutional powers. With a true emergency underway, there's a very tempting opportunity to seize emergency powers... and never let go. Elections would be canceled indefinitely. As Cybereason warned the DHS, martial law is a potential outcome of serious foreign interference.

The United States might remain a constitutional republic, at least in name. But, over the long term it would run as a republic in a state of emergency that requires authoritarian governance. Citizens who complain would be criticized, or maybe even killed through digital doxxing and vigilantism, for suggesting that the emergency is over. The government could take over major media organizations to control information that might affect public safety. Constant digital disinformation would then reinforce authoritarian rule even as digital systems came back online.

This sounds a bit loony, but what I've just described in the last two paragraphs is more or less how Russia has been functioning since the early 2000s. A series of (suspicious) terror incidents precipitated a state of emergency that Putin has exploited to exert dictatorial control over the media and the people. It's not difficult to imagine that he might consider doing the same thing to this country, if he feels threatened.

A military loss

A significant American military setback could cause political changes at home. This has been the pattern throughout history. A government that cannot defend its population is a government that has lost the faith of the governed. The United States has not had such an experience, though one might argue that Watergate qualifies as one. The political changes wrought by the Watergate scandal were a direct outcome of the failed war in Vietnam. It led to major changes in the way the US government functions.

Today, as noted in earlier chapters, the US military is highly vulnerable to cyberattacks. It is possible to imagine a scenario where China, for example, feeling threatened in the South China Sea, decides to take action against the US Navy. By hacking hardware and software throughout the fleet, it causes Naval operations to grind to a halt. Then, through GPS spoofing and digitally tricking Naval detection systems, the Chinese Navy is able to sink the bulk of the US Seventh and Third Fleets.

Fifty-thousand American sailors die. Hawaii and the entire West Coast of the US are now vulnerable to Chinese aggression, with few defenses. US nuclear missiles explode in their siloes due to Chinese hacking. US Air Force planes are grounded due to Chinese firmware hacking. The remainder chase phantom targets or stand down due to impersonation of commanders made possible by the OPM breach.

At the same time, Chinese hackers create mass mayhem on the West Coast by exploiting data they have about Americans and access to devices operating in American homes and businesses. They cause riots and panic, further weakening America's ability to defend itself. China shuts down the entire US military supply and logistics chain by taking over the Chinese-built hardware that runs it. With the military not functioning and mass panic setting in, China takes over the entire Pacific Ocean in a matter of weeks. Hawaii and the Western US could easily become Chinese colonies at this point.

In a scenario like this, what will become of the US government? At the very least, it would be paralyzed by crisis, unable to function normally. The government's incompetent reaction to the COVID-19 pandemic offers a preview of how it might fumble its handling of a military disaster. The backlash from the defeat could trigger epoch-making changes to the government, or even a collapse of the existing system. Alternatively, the trauma could unite us. Again, though, what do we think our currently polarized citizenry will do? Let's not deceive ourselves about the temptations that arise in this sort of situation for cynical, would-be authoritarian leaders.

A contested presidential election

This is the scenario people have been expecting for 2020, as the DHS/Cybereason project suggests. It could be bad, very bad, for the US as a constitutional republic. A thousand different variations exist on the theme of a contested presidential election. "Chaos is the point" of Russian election interference, explained Laura Rosenberger, director of the Alliance for Securing Democracy, which tracks Russian disinformation efforts, in *The New York Times*. "You can imagine many different scenarios. You don't actually have to breach an election system in order to create the public impression that you have."[299]

Alternatively, as former *New York Times* reporter Kurt Eichenwald laid out in a Twitter thread: "So, it's October 25, 2020. Polls show that Trump is going to lose in a landslide. He announces a national emergency/"We have evidence… that millions of illegal aliens are conspiring to vote and undermine our democracy. Therefore, I am declaring a national emergency, suspending elections until this corruption is weeded out. And that's it. People can sue, but his powers are broad enough and his corrupt…/"Judiciary would have to knock it down. And even if it did, there is nothing in the Constitution that says he has to abide by the ruling. Just the courts own interpretations. He can argue that Marbury v Madison does not apply in emergencies, and who can stop him? Ok, now…/….suppose the people take up arms or Does anything but acquiesce. Trump can declare this an insurrection and suspend habeas corpus, meaning they can lock up anyone they want. Lincoln did this and while it stretches the Constitution - the section on habeas corpus appears under…/article one, but never says who has the power to suspend and Trump can simply say, as Lincoln did, I have that power. He also has the authority to shut down the internet or all communications. While technically the electors are supposed to cast ballots, they can't without…/an election victor. And without an electoral vote,

The president cannot be removed./ This is all legal. And does ANYONE believe, given he know this, that Trump won't CONSIDER doing it? A man who publicly stated in 2016 that he might not accept the outcome of the election..."

In his Tweet series, Eichenwald referenced Hitler's rise to power and dissolving of the legislature through the use of national emergencies. He talks about Roosevelt's internment of Japanese Americans in World War II. He alludes to Trump's demonization of the press and his calls to lock up his political opponents. He ended with a warning: "The bottom line: bad stuff could be coming. It is not outside the realm of possibility. It would not even be illegal or unconstitutional. And it will end the great American experiment in democracy. But rich people will do very very well."[300]

Could the US Become a Dictatorship?

This is one of those questions that will probably seem, from the distance of years, to appear ridiculous. Will the US, with 240 years of republican self-government, slide into authoritarianism? All I can say is that the question is more relevant, and less absurd now than it's probably ever been.

Assessing the risk devolves into a contest between overheated anecdotal evidence and a concern that we're ignoring obvious signs of the republic's breakdown. For

Thomas Pepinsky, a professor of government at Cornell University and a nonresident senior fellow in the Foreign Policy Program at the Brookings Institution, "It's hard to deny what's happening here: the support for concentrating federal power in one person is building."

Writing in *Politico*, Pepinsky cited the Pew survey that found 43% of people felt the President could operate more effectively if he did not have to worry about Congress or the courts. He warned that the "hardening split," as he put it, between the parties is becoming a situation where Americans increasingly "label one's political opponents as un-American, disloyal, even treasonous." He added, "Political scientists have a term for what the United States is witnessing right now. It's called 'regime cleavage,' a division within the population marked by conflict about the foundations of the governing system itself—in the American case, our constitutional democracy."[301]

Chinese-made technology is, or should be, part of this analysis. The US is now wired for totalitarian surveillance, mostly by means of Chinese technology that's been eagerly embraced by American consumers. The US government has, as we learned through the Snowden revelations, undertaken many steps to surveil Americans with dubious legal justification. Still, we're not in a surveillance state...

yet. Americans are free to say what they want about the government without fear of reprisal… yet. Could this change?

The potential exists for a totalitarian society, one that completely erases the boundary between public and private life—a situation reminiscent of Orwell's *1984* and Hannah Arrendt's definition of totalitarianism. China offers a model of what could be, with its massive surveillance, facial recognition systems and AI-driven population controls, now being deployed to oppress the Uyghur minority in China.[302]

If there were a suitable crisis to justify it, an American leader who felt he or she had unquestioning support, could easily start abusing the powers of the installed base of Chinese surveillance technology to punish opponents and keep loyalists in line. It's a potentiality that's dangerously simple to realize. All it would take is a decision to do it, coupled with legislative and judicial acquiescence. Once that switch is thrown, it might impossible to reverse the process. The power it confers is too great to be abandoned without a fight.

Conclusion

Predicting the demise of the USA as a constitutional country is a subjective, intellectually risky exercise. The tensions of the current moment make it hard to see what's really happening. And, as is often the case, the causes of national ruptures may not be apparent until it's too late—only evident in retrospect. That said, there's plenty of available evidence to suggest that the danger is real. And, the risks are not about today. Right now, the US has a strong economy and no serious crisis affecting the public. With an economic downturn, perhaps driven by international trade, compounded by destructive foreign cyber meddling, the republic could find itself stressed in ways that we will have trouble comprehending.

CHAPTER
9

Defending the Quantum States of America

The opening of "The Devil and Daniel Webster," a short story by Stephen Vincent Benét, features a New Hampshire farmer encountering the ghost of Daniel Webster, a famous 19th century politician. The ghost thunders, "Neighbor, how stands the Union?" As Benét advised the farmer, "you better answer the Union stands as she stood, rock-bottomed and copper-sheathed, one and indivisible."[303]

How is our union? Today, it doesn't seem so rock-bottomed, copper sheathed, one and indivisible. It feels fractured and hopeless. For certain, there's no one like Daniel

Webster imploring us for it to be so. The USA is in trouble. If we want to save the country, we need to wake up to that fact. Even if the extreme outcomes described so far in this book don't materialize, the country faces serious risks from digital technology, foreign cyberattacks and digital disinformation.

Our Current Cyber Defenses

The US can defend itself against all enemies, even ones that present difficult cyber threats. Every corporate and public sector entity of any size has its own dedicated cybersecurity department. In big companies, there can be hundreds or even thousands of people involved in mitigating the risk of cyberattacks.

The US federal government and state governments are also doing their part, or at least trying to. A lot of people, many of them highly trained, motivated and sworn to defend the United States against all enemies, are working to mitigate the massive cyber risks this society faces. Government entities, standards bodies and private companies are involved in the effort. There are laws and policies similarly aimed at reducing cyber risk. We are doing a lot. It may not be enough, though. And, what we're doing may not be looking at the problem the right way.

Like many of the subjects I cover in this book, a comprehensive review of US government agencies and cyber policies would be overwhelming and distracting. Many online resources can give you in-depth detail if that's what you're looking for. What follows, then, is an overview of the major entities, laws and programs currently in effect.

Federal agencies

The federal government works at cyber defense across a variety of agencies. The National Security Agency (NSA) is among the most prominent, but least well understood. They are involved in intercepting foreign cyberattacks while also engaging in offense cyber programs against our enemies. The NSA has been criticized for keeping cyber vulnerabilities secret so they can use them to attack others—but leaving American computers exposed. They are starting to change this practice. In early 2020, for example, the agency made headlines for notifying Microsoft of a vulnerability in Windows 10, rather than holding the vulnerability back for their own purposes.[304]

The NSA discovery also issued triggered an emergency notification by The Cybersecurity and Infrastructure Security Agency (CISA), to federal agencies to remediate the Windows problem as quickly as possible—a good example of how federal cyber defense can work when everyone is doing their jobs.[305] CISA, which is part of DHS, functions as the main cyber risk advisor to the United States. They focus primarily on securing federal network and digital critical infrastructure, like power plants and dams, but the CISA also finds itself in the lead on many other national cybersecurity efforts. CISA is a new agency, formed in 2018 through the Cybersecurity and Infrastructure Security Agency Act of 2018, which was signed by President Trump. CISA is a continuation of several predecessor agencies, some of which were already operating inside DHS.

The CISA does not work alone. Rather, it has many partners across the government as well as in private industry and the non-profit sector. The agency works closely with industry groups that coordinate security and policies in the electrical power sector, nuclear plants, chemical plants and so forth. This includes the North American Electric Reliability Corporation (NERC). This organization's Critical Infrastructure Protection Standards (NERC-CIP) form the core of countermeasures to protect the American electrical grid.

CISA departments include the National Risk Management Center (NRMC), which is a planning, analysis, and collaboration center for identifying and addressing critical infrastructure risks. They also run the Emergency Communications Division and the United States Computer

Emergency Readiness Teams (US-CERT), which responds to cyber incidents.

The agency is the result of many false starts and a long, tortured path full of bureaucratic infighting. The Aspen Institute's Breanne Deppisch chronicled this history in "DHS Was Finally Getting Serious About Cybersecurity. Then Came Trump,"[306] a *Politico* article well worth reading. She points out, for example, how DHS plans for a bigger cybersecurity project got derailed by the Trump administration's focus on immigration and the subsequent political battle between DHS and the White House over border security.

How is CISA doing? Despite its inauspicious start and struggles, it would seem they are doing their job well. At the time of this writing, there hasn't been a major federal cyber incident since the agency was formed. Nor has there been a serious breach of a critical infrastructure system, at least one that has been disclosed. This is encouraging, though Advanced Persistent Threats (APTs) could still be lurking beneath the surface of government systems.

One CISA program that's drawing praise from industry experts is Continuous Diagnostics and Mitigation (CDM). CDM, which was commissioned by Congress, offers a dynamic approach to fortifying the cybersecurity of government networks and systems. It provides federal departments and agencies with capabilities and tools to conduct automated, on-going assessments.

For Katherine Gronberg, Vice President for Government Affairs at Forescout, a significant player in the device visibility and control field, "CDM is a great program, especially for agencies that are not getting good FISMA scores." Gronberg noted, "Though it's been criticized for moving too slowly, I think anyone in the space would agree that it is still significantly improving the cyber hygiene of federal civilian agencies."

CISA is just one agency. Each federal agency is responsible for establishing cybersecurity standards for itself and entities it works with through the Federal Information Security Management Act (FISMA). This process can be uneven, as GAO reporting has revealed. Then, industry-specific laws that address cybersecurity each have their own agency oversight. The HIPAA law that covers healthcare privacy and cybersecurity is run out of the Department of Health and Human Services (HHS). The Gramm-Leach-Bliley Act, which deals with financial institutions and customer privacy, is managed by the Federal Trade Commission (FTC).

Private corporations receive little or no federal cyber protection. With critical infrastructure companies like power utilities, CISA provides extensive coordination, threat sharing and guidance. For companies outside of critical infrastructure, businesses are entirely self-reliant for cyber defense. This makes sense, because the government cannot possibly protect every American corporation. However, it's extremely difficult for regular companies to fend off nation state actors.

The US Cyber Command

The United States Cyber Command (USCYBERCOM) is one of the DoD's eleven unified commands. Its mandate includes strengthening DoD cyberspace capabilities and supporting both defensive and offensive cyber operations. It was created in 2009, originally as part of the NSA. Their mission statement reads, "USCYBERCOM plans, coordinates, integrates, synchronizes and conducts activities to: direct the operations and defense of specified Department of Defense information networks and; prepare to, and when directed, conduct full spectrum military cyberspace operations in order to enable actions in all domains, ensure US/Allied freedom of action in cyberspace and deny the same to our adversaries."

USCYBERCOM is not the only entity in the US military working on cyber defense and offense. Each branch of the service has its own CISO and cyber operations. USCYBERCOM may play a coordinating role in the work of these other groups. USCYBERCOM is quite small, however, when viewed in the context of the overall US military.

The command has had some successes, such as the successful blocking of Russian social media influencers during the 2018 midterm election. As of late 2019, the Command made an interesting threat—warning Russia it would disrupt the comfortable lives of the ultra-rich oligarchs who dominate the Russian government, if the country interferes in the 2020 US elections.[307] Time will tell whether they will follow through on this threat and if it will have any impact on Russian behavior if they do.

State and local agencies

Most American states have a CISO's office and statewide cybersecurity operations of some kind. The successful state-level ransomware attacks of 2019 showed, however, that states suffer from extensive cyber weakness. According to most observers, these problems are the result of a lack of resources. This issue cuts across many if not all states. They tend not to have enough money or trained personnel to mitigate today's serious security threats.

Solutions are emerging for this problem. These include the use of Managed Security Service Providers (MSSPs), which are outsourced contractors who conduct cybersecurity operations on behalf of their clients. They can be a good option for states, as they relieve much of the day-to-today security monitoring and incident response workloads from overloaded state employees. The state National Guards are also now creating cyber units that can be called into action in the case of emergency or for preventive measures like election security.

Cities are responsible for their own cybersecurity as well. However, except for some of the biggest municipalities, this is a deficient arrangement. The ransomware attacks against Baltimore and Atlanta, to name two of many recent examples, revealed serious vulnerabilities at the local level as well as in local public sector organizations like school districts. This is unavoidable, it seems. It is a huge challenge for cities to recruit and retain skilled cybersecurity people. The budgets for the work may not exist. The Information Technology (IT) staff of local governments are generally too over-stretched and inexperienced to mount a defense against sophisticated attackers.

Laws and standards

Several federal regulations cover cybersecurity. These include HIPPA and Gramm-Leach-Bliley. The most prominent of them, however, is FISMA, which was originally part of the Homeland Security Act of 2002. FISMA "requires the development and implementation of mandatory policies, principles, standards, and guidelines on information security" for government agencies. Any company or public sector entity deals with the federal government must adhere to FISMA.

Like most federal regulations, FISMA is at once complex, sprawling and vague. The specific standards used for FISMA are determined by the National Institute of Standards (NIST). NIST has published various standards and frameworks to enable FISMA compliance. It's way too complicated to get into here in any detail. There are dozens of NIST standards and specialized specifications for data security, encryption and so forth. The essence of FISMA is that it binds all federal agencies to the same standard for cybersecurity. It assigns responsibility for cybersecurity to agency heads and provides accountability through certifications and audits.

However, as GAO reporting has shown, individual agencies may not be doing all they can to stay secure. Critics point out that the FISMA methodology emphasizes planning over the measurement of actual security. Most government security experts feel FISMA has helped the federal government get more secure, but worry that it can risk becoming a checklist rather than a driver of serious security improvement. Observers have also noted that these laws do not cover companies that are critical to the Internet, such as Internet Service Providers, software makers and so forth.

As progress is made in some areas, other parts of the government are clearly lagging. For example, the Office of Personnel Management (OPM), which suffered a catastrophic breach in 2015, has *still* not fully addressed the cybersecurity weaknesses that led to the attack. A 2019 audit found "material weaknesses," in the OPM's the agency's information systems control environment.

For example, as reported in *Federal News Network,* The Inspector General reported that "OPM didn't have a system in place to identify and generate a complete and accurate listing of contractors and their employment status. Additionally, the IG found OPM didn't appropriately provision and de-provision users' access to the network based on their work status."[308] These are exactly the kind of control breakdowns that enable hackers to penetrate networks.

Threat sharing

The government and private industry have gotten a lot better at sharing threat intelligence in recent years. There are now many Information Sharing and Analysis Centers (ISACs) across the US. ISACs are in the business of sharing relevant threat information with interested parties. For instance, if a company in the financial industry discovers a piece of malware, it can share its "signature," or identifying characteristics with

ISACs in the electrical power grid sector and so on. This sharing enables better protection all around.

How Can We Get Better at This?

To get anywhere with improving the USA's vulnerability to cyber threats and related problems with digital disinformation, it will necessary for leadership at many different levels of the government and industry to get more realistic about our current predicament. A lack of clarity, either deliberate or unintentional, is stymying our efforts to develop an effective defense. The Cyberspace Solarium Commission report shows that some elements in the government are trying to overcome this very problem and build a more coherent, resilient national response to cyber threats.

A successful defense will also mean a painful process of self-examination. It requires getting real about who we are and what we need to do about this crisis. What follows here is something of a wish list, one that may never be fulfilled. It's only partly about technology.

Confronting our true national character

In quantum mechanics, a subatomic particle like an electron can be in multiple places at the same time. In a quantum world, there can thus be more than one simultaneous reality. Devising an effective defense of the US from cyber enemies requires us to acknowledge our quantum state when it comes to difficult issues like poverty and race.

What is America, really? Is it a democratic, constitutional republic of States with sovereignty of law and equal rights for everyone… or is it more like Eastern Ukraine, a hate-filled geographical space peopled by white and non-white ethnic groups that hate each other with murderous passions that can be easily aroused through propaganda and demagoguery? It's both at the same time. The US is in a quantum state.

The actual numbers are either scary or reassuring, depending on how you look at them. A Public Policy Polling/ Marist survey in 2017 found that 4% of people who approve of the job Trump is doing as president openly agree with the principles of the Ku Klux Klan. Is that a lot of people or a few? If the poll is correct, it means at least several million Americans accept the Klan's viewpoint on race. It's a small group, in the scheme of things, but it's still big enough to have an impact on public life, especially when their perspective is amplified by a sophisticated digital media machine.

Russian leadership hates America only partly because they think we're encroaching on their geographic sphere of influence. They hate what we stand for, which is equality and self-government by the people and for the people. People like Vladimir Putin recognize the existential threat posed by popular uprisings like the Maidan revolution in Ukraine, where citizens demand a voice in government and an end to rule by self-serving, greedy oligarchs.

How do they respond? They point out that we Americans are not as free as we think we are, that we have our own greedy oligarchs and sham elections—or at least they can make them look like shams. They point out our embarrassing history of racial injustice, telling the truths that we ourselves cannot face: that much of America was built by slave labor on land stolen from Native Americans. Our desire not to "go there" doesn't make this history any less true.

The reality of life and government in the US is far from this ideal. There are many gross distortions of democracy that favor rich people and whites over blacks. Environmental racism is an example, where toxic pollution is zoned into non-white areas. Many historical atrocities have been swept under the rug, such as the 1921 massacre of an estimated 300 African Americans by white mobs in Tulsa, Oklahoma. The number of deaths is an estimate because the episode has never been thoroughly investigated—a very big rug covers up such atrocities, indeed. This dichotomy presents Russia with an outstanding wedge it can use to drive us into internal conflict.

Russia can twist the knives of American history. It can seed doubt and anger by stirring up resentment on both the left and the right. Russia can instigate a march where white men who feel deeply aggrieved about their plight in life can scream "blood and soil" believing that it is their blood and soil that's being stolen from them, when in fact, their ancestors are the ones who shed the blood of others and stole their soil. It may be an impossible mission, but any countermeasures to racially inflammatory disinformation need to include some counter-programming, some admissions of truth as a way to unite people in recognizing how we're being attacked.

Stop being apathetic

The countermeasures also need to confront the uncomfortable reality of American apathy. For the most part, we yawn at shocking corruption in the government, whether it's the "pay for play" of "Clinton Cash" or the self-dealing that characterizes the Trump administration. We can't seem to mount a response to thousands of Latino children being imprisoned at the southern border. We shrug our shoulders at Edward Snowden's revelations of mass surveillance. Our indifference to these situations empowers our enemies. Russia and China see how little we seem to value our democratic system. This apathy emboldens them to overthrow it.

We also seem to be apathetic about clear and present cyber dangers. For instance, when the Senate Report on weapons' systems vulnerabilities came out, there were some murmurings in the press. The story was covered relatively well by outlets like Lawfare Blog[309] and NPR.[310] The *New York Times* covered the report in a story titled "New US Weapons Systems Are a Hackers' Bonanza, Investigators Find." They observed that even the declassified details in the report "painted a terrifying picture of weaknesses in a range of emerging weapons, from new generations of missiles and aircraft to prototypes of new delivery systems for nuclear weapons."[311]

After that? Nothing. A few months later, the Senate Armed Services Committee held a closed-door hearing with some important DoD figures and that was about the last we heard about it. The alarm was raised, but essentially ignored. Why?

Stop making naïve assumptions about America's cyber risk

Whatever we do, or don't do, in the coming years, we should understand that the cyber threats we face are neither going to disappear nor get any less serious. The Cyberspace Solarium Commission report did a good job highlighting the dangers the US is facing from foreign cyber threats. Further, signaling the danger we continue to face, a *New York Times* article co-authored by David Sanger, one of the wisest and most experienced cyber/national security reporters working today, called out Russian efforts to hijack Iranian hacking units to attack the UK and elsewhere in the Middle East—hoping that Iran would take the blame for any damage they caused.

With this in mind, the authors cautioned federal and state officials charged who are tasked with cyber defense for the 2020 election. For Sanger et al, this Russian attack "was a clear message that the next cyberwar was not going to be like the last. The landscape is evolving, and the piggybacking on Iranian networks was an example of what America's election-security officials and experts face as the United States enters what is shaping up to be an ugly campaign season marred by hacking and disinformation."

The article noted that "American defenses have vastly improved in the four years since Russian hackers and trolls mounted a broad campaign to sway the 2016 presidential election," but also shared that "interviews with dozens of officials and experts make clear that many of the vulnerabilities exploited by Moscow in 2016 remain. Most political campaigns are unwilling to spend what it takes to set up effective cyber defenses. Millions of Americans are still primed to swallow fake news. And those charged with protecting American elections face the same central challenge they did four years ago: to spot and head off any attack before it can disrupt voting or sow doubts about the outcome."[312]

Having spoken to hundreds of people in the industry about these issues, it is evident that the biggest problem has to do with poor assumptions. These are experienced, thoughtful and intelligent people. They are concerned about the cyber risks facing the US. They want to help. In some cases, they make their livings helping the US solve cyber defense problems. However, as human beings, they may be failing to see the implicit assumptions they're bringing to the issue. These include:

- **That the US will continue to function despite being paralyzed**—There's a tendency to downplay or not even acknowledge that a country seized by panic and mayhem may not have a functioning government. Certainly, at the state and local level, where ransomware has demonstrated that it can bring government and emergency services to a halt, there may be no working institutions or organizations to mount a meaningful response to the attacks.

- **That the US government will want to solve the problem**—Given the polarized, overheated state of politics, one side or another might simply leverage a cyber crisis for its own ends, rather than solve the problem. This might involve prolonging or exacerbating the problem for political reasons or even using the crisis to justify the imposition of non-democratic (and long-term/permanent) emergency measures.

- **That people engaged in cyber response will be able to do their jobs or communicate with one another**—Cyber kill chains and incident response plans invariably include communications between teams and people by email, website interfaces, text and phone. If communication networks are compromised and people are missing or dead, the response will be impaired.

- **That foreign cyberattacks are different from military attacks on sovereign territory**—Somehow, perhaps because the attacks are virtual, it's hard for policy-makers to call out cyberattacks as foreign attacks on American soil. When our enemies enter our networks and do damage to our corporate, government or military infrastructure, it's no different from them launching missiles at us. In the OPM breach, for example, perhaps if we saw Chinese soldiers breaking down the doors of the agency and ransacking computer systems in physical space, we might feel differently about their theft of 23 million government personnel files.

- **That military and civilian cyberattacks are separate and unrelated**—Civilian infrastructure has military significance. A ransomware attack on San Diego, for example, could affect the Navy's ability to function, given that San Diego is one of the Navy's most important installations. More broadly, a deeply crippling attack on the military supply chain, which comprises multiple civilian targets, could impede American military defense. The US once understood this, as they carried out the strategic bombing of Germany's industrial base in World War II to stop the Wehrmacht from advancing. Or, in the case of the French city of St. Nazaire Effect, which housed the German U-Boat command, allied bombers leveled the entire city and its railway tracks to stop resupply and repair of an enemy that was sinking thousands of tons of allied shipping for the war. The same kind of attack can take place today against the US through cyber techniques.

- **That the intent to respond is a response**—The government tends to confuse taking steps to defend itself with having a working cyber defense. This may be more a problem of bureaucracy and budgets than individual incompetence. However, it's one thing to mandate a security framework or pass a law like FISMA and quite another to be able to detect and respond to cyberattacks. The 2019 report, *Federal Cybersecurity: America's Data at Risk*, produced by the Senate Permanent Subcommittee on Investigations, which identified seven government poorly defended agencies, underscores this deficiency.

Similarly, the DoD's response to the serious report on weapons' system cyber vulnerabilities has been to mandate that defense contractors adhere to the NIST 800-171 standard. This is a wise move, but a realist will see that it's one with limited impact. As Assistant Secretary of the Air Force, Dr. William LaPlante, explained the Committee, told the Senate Committee, "There is no measure or target for outcomes associated with implementation of the 800-171 standard. For instance, was less data lost? While standards may have the potential to improve performance above a baseline level, they quickly lag behind evolving operating environments and emerging technologies. Most importantly, they quickly become the target of our adversaries, who familiarize themselves with our standards and look for seams they can compromise."[313]

In other words, the standard can only do so much, even if it's perfectly implemented. And, as we know from a constant drip of serious cyberattacks on defense contractors, compliance is not consistent.

- **That global shared networks reduce the relevancy of the traditional nation state**—The enthusiasm for the Internet and the global connectivity of the Internet, which is exciting and real, serves to obscure the unchanging reality that humanity is still divided into nation states. We may be able to browse pictures of Russian artwork in a Moscow museum online, and that's cool, but Russia is still a separate country, with its own government and national goals. The fact that the Internet is an American invention, and computing in general has been so US-centric for so long, may lead us to assume that everyone else in the world has the same view of the world as a place that loves America. It may be hard to imagine that others could use our invention to hurt us.

Stop taking refuge in ass-covering rabbit holes

Getting beyond the limits of such assumptions will be a monumental challenge. In my view, the people involved have the brains and aptitude to pull this off. However, it will require the discovery of a quality that seems to be in short supply: leadership. To this point, Michael Dent, the Chief Information Security Officer (CISO) of Fairfax County, Virginia made the following comment on the cyber disasters of 2019. "*When I peel back the onion, there is one commonality among all of them–LEADERSHIP FAILURE. This is where we turn-on the light and the roaches disperse, hide, build excuses, start pointing fingers and ultimately punish those that warned leadership in the first place…*"[314]

Being situated in a place where a lot of US government digital assets are housed, Dent is in a good position to see the nature of the problem. What's driving the excuse-building, finger-pointing exercises he sees? It's human nature, of course, but it also bespeaks a certain kind of counter-productive government mindset.

I call it the "ass-covering rabbit hole" phenomenon. When confronted with clear cyber risk, government officials and cyber experts, many of whom come from government service, either cannot or will not acknowledge it. Media-trained in the art of obfuscation, they shift conversations about America's real vulnerabilities into semantic and evidentiary game playing. Techniques include changing the subject or burrowing into a deliberately narrow discussion that runs out the clock.

For example, I once asked a former senior US foreign policy official why there hadn't been greater accountability for defense contractors whose negligence led to the theft of the F-35's digital plans. Even though this gentleman had been out of government service for years and was not being asked to speak on the record, he still retreated into non-denial denial mode, saying "They only stole unclassified data and the Chinese replica of the F-35 is a substandard aircraft."

Or, when reviewing the GAO report that described major cyber defense weaknesses in government agencies with a woman who had previously served in government cybersecurity roles, I could not get her to give a clear opinion on whether any of these weaknesses might have led to data breaches at federal agencies. "You can't make that assumption!" she cautioned me. No, one surely can assume that unpatched government systems, which were readily discoverable by the GAO, have been exploited by the Chinese or Russians.

In another case, I tried to get a meaningful comment out of the DoD in the aftermath of the Polar fitness tracker episode. Recall that six months after the Strava tracker incident, in which fitness trackers exposed sensitive troop locations in Afghanistan, the Polar Fitness mobile phone app was revealed to be vulnerable to hacking that showed the movements and identities of American military personnel and intelligence officers around the world.

Curious about why the DoD had not been able to stop military personnel from using the insecure and consumer-grade Polar app while on military business, I asked the DoD spokesperson if any security policies had changed or been developed for military personnel regarding personal fitness trackers or other tracking devices.

He responded, "With regards to the Polar fitness App: We are aware of the potential impacts of devices that collect and report personal and locational data. Recent data releases emphasize the need for situational awareness when members of the military share personal information." He then added, "Annual training for all DOD personnel recommends limiting public profiles on the internet, including personal social media accounts. Operational security requirements provide further guidance for military personnel supporting operations around the world."

Finally, he noted, "The Under Secretary for Defense Intelligence is writing guidance to emphasize the risks of using global positioning system-enabled devices and to direct components to ensure local operations security policies are adequate. DOD is constantly reviewing our force protection methods to determine if any additional training or guidance is required in order to ensure the continued safety of DOD personnel at home and abroad."

A friend of mine who graduated from West Point and served in the Army characterized the spokesperson's response to me as "rectal persiflage." The DoD finally set down firm guidelines restricting the use of fitness trackers and location tracking devices by troops deployed in sensitive areas in August of 2018.[315] It took eight months. And, this was five years after the Defense Science Board's stark warning on the cyber vulnerability of the US military and ten years after the theft of the F-35 designs by the Chinese. Evidently, the new rules didn't govern TikTok, use. That required a separate memo yet another year later.

Now, I realize that they're media relations people and I'm the media, so they're going to give me the "persiflage." Still, the dialogue is emblematic of how difficult it can be to get anyone to discuss these issues in a direct, productive way.

Have a sense of history

Cyberattacks and disinformation directed at the US are simply new fronts in old conflicts. It's counter-productive to view them as different just because the technology is recent. Setting aside gross ignorance, it's a problem when even seemingly well-informed Americans do not understand the long historical roots of current cyber animosity.

There's tendency to view attacks in a narrow time context, e.g. we got attacked last year. Let's deal with that. Yes, we do need to respond to the most recent attacks, but we will only arrive at a truly effective solution if we grasp how Russian attacks, for example, are based on resentments that started to form in the Cold War or even earlier.

The US has been in an antagonistic relationship with Russia since 1917. The US supported the "Whites" who fought against the "Red" Bolsheviks in the Russian civil war of 1917-1921. After the Bolsheviks founded the Soviet Union, their resentment of American interference lingered. That may seem like ancient history, but Russians have long memories.

After a brief alliance of convenience between 1941 and 1945 during World War II, the US and USSR sank into the bitter, global geopolitical struggle of the Cold War. This lasted from 1946 to 1989. We were arch enemies, spying on each other relentlessly, constantly threatening one another with nuclear annihilation. The US had nuclear-armed B-52 bombers in the air, 24 hours a day, 365 days a year for decades. The Russians were aware of this. They haven't forgotten what that threat felt like.

Then, after the collapse of the Soviet Union, which was partly due to US economic warfare, Russia perceived that the US was trying to keep Russia in an inferior geopolitical position. A period of attempted cooperation ended in the early 2000s with both countries retreating into a familiar mode of mutual distrust. So, the attacks we're experiencing now are just a digital version of some really old patterns.

China has similar deeply rooted suspicions and resentments of the US. The Cold War was also a conflict between the US and China. It may not have been as elaborate and threatening as the US-USSR conflict, but it was serious. China is angry that the US supported the Nationalists in the Chinese civil war of 1945-1949. They blame the US for the rise and success of an independent Taiwan.

To further understand American cyber vulnerability in the historical context, it's worth remembering how the United States even came into existence in the first place. White Europeans came to North and South America in the 15th century. The continents were settled at that time by hundreds of Native American nations, some of them advanced empires like the Aztecs. What happened to them?

The Aztecs and millions of Native American people died from diseases brought by the European invaders. This is the only way that a few thousand sickly Europeans, having just come out of an arduous ocean crossing, could conquer empires and nations with millions of citizens. We got them sick and they died. It was one of the greatest tragedies in the history of humanity.

We have a similar vulnerability. We risk becoming digitally "sick." Just as bacteria from Spain made the great Aztec armies die off, so too can Chinese and Russian digital viruses make our great armies sick. Our technology is exposed to foreign infection. Americans are vulnerable to a stealthy plague, a digital illness that can wipe us out despite our military strength and destroy our system of government.

Conclusion

It's hard to feel safe at this moment in time. The country is under attack. The consequences of a failure to defend ourselves may be severe in the extreme. Professionals are hard at work trying to solve the various problems, though the inevitably slow pace of government operations gets in the way. Some of the issues are not technical in nature, but rather have to do with an inability or unwillingness to see what is truly happening.

CHAPTER 10

Making the Machines (and Ourselves) Smarter

C an there be hope, after all this? Can we solve this crisis without triggering World War III or a meltdown of the US as we know it? There are reasons to be hopeful that the US will rise to the occasion and develop more effective defenses against the aggression we now face. It will take effort, but more importantly, a strong cyber defense will require a change in thinking. To defend ourselves, we have to rethink our approach to computing and networks as well as the global/national paradox of cyberspace. Some of the solutions are technical, but many are not.

This is the time to take action. We are at the start of a historical cycle, with much more and much worse to come. At the very start of 2020, for example, the U.N. General Assembly approved a resolution for Russia to begin drafting a new global treaty combatting cybercrime.[316] It sounds like a joke, but it's not a laughing matter. The country that is arguably the world's number one cybercrime offender wants to write the treaty dealing with the problem.

A week later, Russia was caught hacking into Burisma, the Ukrainian company at the center of the Trump impeachment scandal.[317] This revelation highlighted Russia's interest in continuing to influence American politics from afar. The same week, Vladimir Putin arranged for changes to the Russian constitution that would allow him to retain power after his official term expires.[318] Whatever we do about all of this, we better get onto it as soon as possible, because trouble is coming.

There are No Easy Solutions

The memorable maxim, "For every problem there is a solution that is simple, neat—and wrong" has been attributed to Mark Twain as well as H. L. Mencken, and Peter Drucker. It's a wise concept to keep in mind if we're contemplating solving the serious, but highly complex cyber risks facing the

US. There are no easy solutions. In some cases, incremental improvements may be all that is possible. For other problems, there may be no solutions at all. However, it's imperative that we try to find solutions. To avoid the issue is to concede defeat prematurely.

The following ideas comprise a fair amount of wishful thinking. They're somewhat contrarian, too, to the prevailing ethos of openness that dominates so much of the thinking in the American technology industry and cyber policy establishment. Each idea is "impossible," according to many experts, on the levels of engineering, industry enthusiasm and government incentives. Each comes with its own set of negative consequences, which will have to be managed. Yet, as we know, almost anything is possible if we want it to happen. I've deliberately tried to stay away from specific technical prescriptions like "endpoint management" or "federated identity management," as these change over time and lead to distracting debates on the merits of one technology over another.

Other books have gone into great depth on some of these issues. Professor Martin Libicki, who teaches at the US Naval Academy, lays out an extensive series of cyber policy positions related to increasing US national security in his book, "Cyberspace in Peace and War." It's a worthwhile read, in part because Libicki shows how complicated matters can get when trying to solve problems that affect the American public, the government and industry.

First, Take This More Seriously. Then, Cool Down.

The republic is in danger. We are at war, give or take, with two other nations. They certainly refer to their operations against the US as war. Why should we pretend it isn't happening? We need to "own it," as those younger than me might say.

People can disagree on the level of risk and the potential consequences of cyberattacks and foreign disinformation, but the danger is present no matter how much we want to wish it away. Our political leaders and media outlets would be wise to start taking this more seriously. This may not be a winning proposition. It's not a sexy story, at least not yet.

While cyberattacks and cybersecurity do get government and media attention, there is little awareness in these circles of the collective threat. It's certainly possible, however, that events will force the issue to the forefront. A destructive, politically-motivated foreign cyberattack could press politicians into action, or at least acknowledging the danger.

Then, to get anywhere in solving the complex problems presented by foreign cyberattacks and disinformation, it will be necessary to take a step back from the current environment of hype and spin that surrounds so many political issues. We need to cool down the rhetoric and gamesmanship to address the difficult questions this issue raises. This may be the most wishful of all the wishful thinking expressed here. Serious American problems deserve serious attention. The current debate and decision making mode in government is not suited to solving complex problems with much wisdom.

Attack the Root Causes of the Problem

Having covered American cyber vulnerability for years, my one categorical takeaway is that the problem will not be solved until digital technology itself becomes more innately secure. The current approach of spending 124 billion dollars a year[319] to keep up with cyber attackers, who are always a step or two ahead of us, cannot be sustained. We are driving ourselves crazy, fighting a losing battle to secure technology that was never designed for security in the first place.

Part of the challenge here involves confronting the fact that we face multiple, interlocking problems. Look no further than the dozens of cybersecurity solution categories to see the variety in threats and vulnerabilities. However, the whole array of risks, from application security to network security, data security and endpoint security flow from a small number of root causes. In my view, the following comprise the root causes of America's current state of cyber weakness:

- Open system architectures that produce endlessly reprogrammable "Turing Machines"—making it easy for attackers to install malware and hijack systems

- Anonymity on networks—enabling hackers to hide themselves when attacking by exploiting misplaced or deficiently-controlled trust

- Open-source software—giving attackers unlimited knowledge of the cores of our digital assets

- Open global networks—allowing foreign hackers to penetrate the US with relative ease

Do I sense an eye roll here? Each of these risk factors is also a driver of explosive growth and most of the vast benefits of modern digital technology. Whole industries and huge fortunes have grown out of open system architecture, network anonymity, open-source software and open global networks. There wouldn't even be an Internet or the world of digital devices we have without them.

Yes, to criticize these fundamentals of a multi-trillion-dollar industry seems foolish and naïve. It is, but in my view, if there is to be any hope of defending the US as a constitutional republic against foreign digital interference, we are going to have to get out ahead of the problems these technology paradigms are causing.

The media and political aspects of America's digital risk are a different story, though these, too, have root causes. I will deal with those, as well. The solutions are less clear-cut, however.

Make Computers and Other Digital Devices More Intrinsically Secure

Turing Machines are endlessly reprogrammable and thus endlessly hackable. One solution is to replace them, where possible, with devices that have programming embedded in their silicon. This is known in the industry as "hardware-based security." For example, a hardware appliance can open web URLs and let any malware on the site "detonate" inside the appliance. The appliance, which runs no reprogrammable code of its own, then displays a visual replica of the website to the end user. This process reduces the odds of web-borne malware getting onto the end user's device.

Hardware-based security isn't ideal for every use case. But, even with its limited uses, it can still help bolster security. And, there may be more uses for hardware-based, non-Turing Machines than we think. The tech industry is very enamored with the Turing Machine model, because they roll out new features and software upgrades without requiring people to get new hardware.

However, if security becomes a serious enough issue, people may be open to trading flexibility for security. For instance, if one could buy a smartphone with fixed functionality like email, chat and web browsing and a few favorite apps—pre-installed on the silicon and unchangeable for the life of the device, that might be a satisfactory and infinitely more secure technology than what's on the market today.

These types of solutions can be updated in their firmware. This procedure can be made fairly secure by adding cryptographic "certificates" that authenticate the entity delivering the new firmware. As with anything in cybersecurity, there is no 100% bulletproof solution, but certificates reduce the risk of malicious firmware updates on hardware-based devices.

The Cyberspace Solarium Commission report contains a further potential solution. It recommends the establishment of a National Cybersecurity and Certification and Labeling Authority.[320] This entity could function like the Underwriters Laboratory (UL) that attests to the safety of electrical appliances. It won't solve all the problems, but it would be a good step toward ensuring security is increasingly built into digital devices.

Embrace Proprietary Code and Closed Architectures

As the comparative security success of the iPhone has shown, erecting barriers to hackers in the forms of proprietary code, closed architectures and license-based software development can impede malicious actors quite effectively. Proprietary code of course has flaws and hidden exploits, but at least the source code isn't available to all the crooks and spies who want to use it to destroy us. Keeping things hidden can be good for security. It's time to make the hacker's job a little harder.

Trust Less, Or Not at All

American networks would benefit from greater use of the "Zero Trust" model of security. In this approach, no user or device is trusted for any use at any time—until they are approved for whatever limited access rights they need. That way, if a malicious actor breaks into a corporate or government network, he or she can do nothing: no installation of new software (malware), no access to databases, no exporting of files (data breaches), no access to system configuration interfaces and so on. The user must request permission from an administrator to be able to perform any such task.

The difficulty that arises here, however, is administrative overload. If millions of users need to request permission to do every single one of their tasks, every time, nothing will work.

The solution is to automate trust assignments using AI, so users with familiar patterns of use can be trusted by default, whereas unknown or odd-behaving users will have to ask for permission. It is of course possible for hackers to hack the AI algorithm and trick their way into being trusted, but again the point here is to make things harder for them. Zero Trust is another layer of defense that stops hackers from enjoying an "all you can eat" buffet of data theft and system disruption if they manage to penetrate the network perimeter.

Reduce Anonymity on Networks

Network anonymity is a problem for American national security. While we require passports, border control and customs for foreign individuals entering the US, a web visitor can cross our borders at will, anonymously. Network anonymity enables Chinese hackers to steal enormous amounts of data from American corporations and defense contractors. Taking advantage of network anonymity, Russian trolls in St. Petersburg can easily surface in the US and post racist comments on neo-Nazi websites, and on and on.

What can be done about this? With the caveat that there no easy solutions, one approach to rectifying this security deficit would be to require Internet users to authenticate their identities before being granted access. This is already starting to happen. For instance, if you want to post a political ad on Facebook, they need to see your driver's license so they can validate that you are who you say you are. At the same time, Facebook has announced that it will continue to allow political ads that tell lies.[321]

User identity authentication also happens on the payment processing parts of the Internet. You need to prove who you are before you buy something online or do a banking transaction. This often involves a process called Multi-Factor Authentication (MFA), which might require you to enter a one-time use password you get in a text message before being allowed to pay with a credit card online.

Some companies are pioneering technologies that can authenticate you through implicit data points like the location and behavior of your smartphone, e.g. if someone lives in New York, but it looks as if their phone is in Moscow, that might suggest a hack is going on. The system can then send security questions to figure out if the phone's user is on a trip or has been compromised.

Ultimately, a push to reduce network anonymity may result in the segmentation of the American Internet. There could be two networks, one for public, anonymous traffic, and another for business and government. This already exists, to some extent, with federal and military networks. Without proper authentication, a user would not be allowed onto the restricted networks. Advances in unique, cryptographic identification, along with technologies like Blockchain, can help make this a reality.

Get Better at Filtering International Network Traffic

Further to the notion of preventing anonymous foreign visitors entering the US digitally whenever they please, it would be a good idea to get better at filtering international network traffic. Of all the "that's impossible" ideas presented here, this is one of the most impossible. Or, so I've been told. That said, the Internet is a work of engineering. It can be re-engineered if the fate of the nation depends on it.

The US is the hub of many international network connections and fiber optic cables that carry Internet traffic. They're run by the big phone companies. The job of filtering out non-US traffic is evidently a massive undertaking. But, it's necessary, in my view, if we want better security. Certainly, Russia and China have realized this. They've both implemented mechanisms to screen out foreign Internet traffic.

Russia's and China's moves have drawn howls of protest from many quarters. Claims of censorship and repressing abound. They're all true. Russia and China want to shut off external Internet traffic because it interferes with their authoritarian systems of government. However, their "national firewalls" also make them far more secure from external hackers. We have no such protection. It should be possible, in this great and innovative country of ours, to devise a network architecture that stops suspicious foreign network traffic while preserving freedom of information and a free press.

Make Data Less Valuable

Hackers steal data because it has value to them. They can sell it or use it for espionage purposes. One countermeasure, therefore, is to make data less valuable. For example, imagine if social securities were binary in nature—that they consisted of a public number and a private PIN, known only by the holder of the number. Then, the number itself is pretty much useless to a hacker. Stealing someone's identity becomes a lot harder as a result.

Alternatively, given the costs of data breaches, businesses and government agencies may decide to store less data on their systems. Instead, there could be a system of highly secure data repositories that companies and the government accessed when they need a piece of information, like someone's birthday or driver's license number. Such a scheme would concentrate sensitive data in the most protected places, reducing the incidence of massive data breaches.

The Power of Automation and AI

Aside from a lack of will or interest, the biggest hurdles to solving these root causes of national cyber insecurity are deficits of money and talent. The cybersecurity industry is experiencing a huge labor shortage. And, even if there were abundant people, doing cybersecurity properly takes a lot of money. To address these twin challenges, many innovative companies are using AI and Machine Learning to take over some of the tasks previously done by people. There are also more automated solutions for managing security and responding to security incidents. It's still early in this cycle of technology development, but it's very promising.

Stop Letting Our Enemy Build Our Technology

The United States, its people, government and corporations will never be truly secure until we stop letting Chinese businesses build our technology. There is ample evidence that the Chinese Communist Party, which either controls or influences the major electronics manufacturers in China, is exploiting the opportunity to create products that spy on Americans or contain built-security flaws. This has to stop. At the very least, the technology used by the military and government needs to be 100% American-made.

Impossible! That's what everyone says. And, the truth is that it will be extremely difficult to bring manufacturing of digital devices back to the US. The cheap shot one hears the most often runs along the lines of, "What, you want to pay $2000 for an American-made iPhone?" Let's deconstruct this idiotic thought process to see what might be actually feasible.

Yes, making products in the US, where people (hopefully) get paid a living wage and don't need nets to keep them from killing themselves in the factory, will result in more costly products. However, in the military use case, for example, it is quite easy to imagine a $2000 iPhone becoming just another piece of military gear. They're already buying expensive equipment. For consumer uses, a $2000 iPhone might be viable if it were offered for a longer contract. So, it will cost $50 a month for its 4-year life. Upgrades won't be as frequent. Can people live with that? Perhaps not.

Certainly, the US should not let Chinese made technology be used in our telecommunications infrastructure. We are at war with China, like it or not—certainly enough of a war that it would be foolish in the extreme to let them build our entire communication system. The government is wisely now trying to prevent Huawei products from becoming the core of the new 5G wireless infrastructure. The US Senate, in a bipartisan move, is trying to pass legislation to fund $1 billion in R&D for an American 5G alternative.[322]

The CCP is sufficiently influential, however, that high-level people in the US government advocate for companies like Huawei, despite the threat it poses. In January, 2020, the DoD blocked a Commerce Department directive aimed at stopping Huawei from buying American-made technology. The DoD felt it was not in the interest of American industry to be constrained in this way. Senators Ben Sasse, Tom Cotton and Marco Rubio complained, writing "Huawei is an arm of the Chinese Communist Party and should be treated as such." They then added, "It is difficult to imagine that, at the height of the Cold War, the Department of Defense would condone American companies contracting with KGB subsidiaries because Moscow offered a discount."[323]

Making more tech products in the US solves another problem, that's related to the entire digital downfall scenario. It will put Americans to work in high-tech, high-paying industries. The downward economic spiral so many Americans find themselves in today, post-China shock, could be interrupted in many cases by opportunities to work in American manufacturing of digital devices.

Equate Foreign Cyberattacks to Invasions of Sovereign Territory

Mitigating foreign cyberattacks will also require changes in policy and perspective. To date, digital incursions into foreign territory have been treated as routine espionage—subject to the vague but still important "rules" that cover such conduct. As long as no one is getting hurt, it goes on without much comment. If we hack Russia, it complains but doesn't take much official action. We are the same. This does not have to be the case, however.

Given the scale of the damage the US is now facing with foreign cyberattacks, it might be time to revisit the old rules. What if a foreign cyberattack was categorized as an unlawful invasion of sovereign territory? This is the way an incident at sea would be approached. If an American Navy ship enters China's territorial waters, that's a big incident in the making. At a minimum, it would result in a warning, with China having every right to make a stink about it. Digital transgressions of national boundaries could be in the same category. This would give countries the right to protest hacking more formally if they so chose.

Negotiate and Retaliate

The US has the right to request that other nations not invade its digital networks. Other countries should have the same right. It sounds silly, but we need to ask Russia and China to stop hacking us. We would have to make a similar pledge. But, why not? Or, perhaps it's time for a cyber treaty that defines what is and isn't acceptable. Existing laws are clearly not adequate.

The UN is taking on this issue. In late 2019, 114 NGOs and nearly 100 governments met at the UN to discuss cyber norms, rules, and principles for state behavior.[324] Where that process will lead is anyone's guess. However, it is a step in the right direction to at least have countries talking about changing rules of cyberspace and national security.

We can also retaliate more, or differently, in reaction to devastating foreign cyberattacks. This need not be the "hacking back" response that is so often discussed. The US has many ways to make life very uncomfortable for the people who run Russia and China. We know what to do and how to do it. The recent threat made by the US Cyber Command to respond to election interference by disrupting the lives of the rich oligarchs who keep Putin in power is a welcome step in this direction.

To make this work, the US might have to limit its own cyber aggression. It's a tricky picture, for sure, but one that good leadership can go a long way to addressing. (Wishing for good leadership may yet be the most naïve idea in this entire book, but I digress.)

Establish Better Media Reporting Standards

The news media has a role to play in mitigating the digital disinformation that's affecting American politics and life in general. It's hard to know what this might look like, exactly, but one idea is for the major news media organizations to develop shared standards about how they will report on social media feeds, Twitter, web videos and the like. If there could be an agreement not to publish uncorroborated web material or unverified, suspected deepfake videos, that could cool the impact of false stories planted for destructive effect.

Turn Off Foreign Digital Propaganda

It's time to turn off RT. Is it so important for cable and satellite operators to carry this channel? It pipes toxic Russian propaganda right into the US. If our leaders can accept that we are in war with Russia, minus the shooting, they might see their way to inducing the TV companies to push the enemy off our airwaves. So, too, should media figures and other public influencers on the CCP payroll be publicly identified so their commentary can be viewed in the proper context.

Get American Corporations to Rethink Their Reasons for Being

One of the non-technical root causes of the current crisis has to do with the changing nature of American businesses. With globalization, American corporations have become more international. That's good business, but what seems to have gotten lost is a corporate sense of duty to America.

When China steals American trade secrets, American corporations are afraid to speak up. They don't want to miss out on opportunities to do business in China. As Assistant Attorney General John Carlin told Lesley Stahl on *60 Minutes*, "Getting CEOs from those companies to talk is nearly impossible because most of them still have business in China and don't want to be cut out of its huge market."[325]

John A. Burtka IV, executive director of *The American Conservative*, put the matter across more forcefully in a *Washington Post* opinion piece titled "American Businesses at a Crossroads." He wrote, "The United States is at a crossroads. We live in a country where the titans of industry have betrayed the national interest in search of profits from a communist dictatorship."

Burtka was particularly incensed at American businesses that acceded to Chinese demands for censorship. He added, "Companies that have prospered on account of the United States' political and economic freedom, infrastructure, and military protection, now limit the speech of Americans and others seeking freedom to appease their minders and masters in Beijing. The consequence of this? The abolition of national sovereignty, individual liberty and the consent of the governed."[326]

Then Again, the Problem May at Least Partly Solve Itself

One additional possibility is that a certain amount of the problem may solve itself. This may seem like a very remote proposition given the scale of the trouble, but market forces can exert greater influence on technology companies than government regulations. For example, if consumers become more passionate about security, they may demand more secure mobile devices and PCs. Whoever makes the most secure phone will then proceed to dominate the market. Insecure players will go out of business.

The same thing could happen for network equipment, computer servers and the like. Similarly, if corporate customers demand that there be a secure Internet, it might be remarkable how many "impossible" engineering feats could occur in a short period of time.

Regarding disinformation, the apparently hopeless cause of making the public more savvy could also come to pass. Younger voters are already much more circumspect about the veracity of online information than their elders.

CONCLUSION

Digital Hope

The US appears to be in danger of falling apart, on the verge of a political disaster that's partly of our own making. The digital technology we rely on is insecure, exposing our society to disruption that could easily spiral into politically disruptive mayhem. This risk is greatly aggravated by our geopolitical adversaries. Our military is also vulnerable in ways that its commanders are only starting to grasp. It's happening in plain sight, but still hard to see.

There are solutions. They won't be easy or simple. In some cases, they may be incomplete. But, it is possible to reduce America's cyber weaknesses and their political ramifications. It will involve attacking the root causes of the problem—making the technology itself inherently more secure. Some of the solutions have nothing to do with technology, but rather with policy and diplomacy.

It may yet even be possible to remain hopeful in this moment of impending crisis. For one thing, the crisis hasn't fully arrived yet, and it may never come to pass. However, even if we avert a crisis today, we need to accept that the world has changed. In order to be safe, the US will have to change the way it deals with other countries around cybersecurity and cyberwar issues.

The technology that presents us with so much risk also carries such great promise. If we can manage digital technology responsibly, it will lead us to a period of greater health and knowledge as a nation. It can be an engine of our economic prosperity.

Ultimately, whether we succumb to a digital downfall or rise to the occasion and address the risks we face will be a matter of leadership. The current era is distinguished by a striking lack of it, on virtually all fronts. The country needs leaders who can return us to a vision of America that is based on truth and unity, despite the so many forces pulling in the other direction. But, the country has overcome major problems in the past, so perhaps we will be able to do this again today.

If we move forward with purpose and speed, we can overcome the problems we face and mitigate the grave risks endangering the country. Digital technology can even lead us to greater security, if we treat it the right way. There is much work to be done.

BIBLIOGRAPHY

Beebe, George S. *The Russia Trap*. Thomas Dunne Books, 2019

Carr, Nicholas. *The Shallows*. Atlantic Books, 2010

Kaplan, Fred. *Dark Territory: The Secret History of Cyber War*. Simon & Schuster, 2016

Kershaw, Ian. *The Global Age: Europe 1950-2017*. Viking, 2019

Koppel, Ted. *Lights Out*. Broadway Books, 2015

Libicki, Martin. *Cyberspace in Peace and War*. Naval Institute Press, 2016

Menn, Joseph. *The Cult of the Dead Cow*. PublicAffairs, 2019

Menn, Joseph. *Fatal System Error*. PublicAffairs, 2010

Mitnick, Kevin and Simon, William. *The Art of Intrusion*. Wiley, 2005

O'Connor, Cailin and Weatherall, James Owen. *The Misinformation Age*. Yale University Press, 2020

Postman, Neil. *Amusing Ourselves to Death.* Methuen Pub Ltd. 1987

Quade, Phil, *The Digital Big Bang.* Wiley, 2019

Roll, David. *George Marshall: Defender of the Republic.* Dutton Caliber, 2019

Sanger, David. *The Perfect Weapon.* Crown 2018

Sciutto, Jim. *The Shadow War.* HarperCollins, 2019

Snyder, Timothy. *The Road to Unfreedom.* Tim Duggan Books, 2018

Snyder, Timothy. *On Tyranny.* Tim Duggan Books, 2017

Spalding, Robert. *Stealth war: how China took over while America's elite slept.* Portfolio/Penguin, 2019

Webb, Maureen. *Coding Democracy: How Hackers Are Disrupting Power, Surveillance, and Authoritarianism.* The MIT Press, 2020

INDEX

ENDNOTES

1   Barbie L. Nadeau, "Trump Cites 'Civil War Like Fracture' Threat If He's Removed," The Daily Beast, September 30, 2019, https://www.thedailybeast.com/donald-trump-cites-pastor-robert-jeffress-threat-of-civil-war-like-fracture-if-hes-removed-in-impeachment

2   Tal Axelrod, "Voters Believe US Two-Thirds of the Way to 'Edge of a Civil War': Poll," The Hill, October 23, 2019, https://thehill.com/homenews/news/467143-voters-believe-us-two-thirds-of-the-way-to-edge-of-a-civil-war-poll.

3   David Von Drehle, "Vladimir Putin's virus: How the Russian president has infected our national trust," Washington Post, March 2, 2020, https://www.washingtonpost.com/opinions/2020/03/02/did-vladimir-putin-turn-america-itself/?arc404=true

4   Chauncey Devega, "Soldiers of the boogaloo: David Neiwert on the far right's plans for a new civil war," Salon.com, May 18, 2020 https://www.salon.com/2020/05/18/soldiers-of-the-boogaloo-david-neiwert-on-the-far-rights-plans-for-a-new-civil-war/

5   United States Senate, "(U)Report of the Select Committee on Intelligence United States Senate on Russian Active Measures Campaigns and Interference in the 2016 U.S. Election: Volume 2: Russia's Use of Social Media with Additional Views" (116th Congress: Report 116-xx, 2019), https://www.intelligence.senate.gov/sites/default/files/documents/Report_Volume2.pdf.

6   United States Senate, "(U)Report of the Select Committee on Intelligence United States Senate on Russian Active Measures Campaigns and Interference in the 2016 U.S. Election"

7   Dennis C. Blair and Keith Alexander, "China's Intellectual Property Theft Must Stop," New York Times, August 15, 2017 https://www.nytimes.com/2017/08/15/opinion/china-us-intellectual-property-trump.html.

8   Jeremiah Jacques, "You're on File: China Building Huge Espionage Database on Americans," The Trumpet, January 17 https://www.thetrumpet.com/18425-youre-on-file-china-building-huge-espionage-database-on-americans

9   David Lumb, "Chinese Hackers Stole Undersea Warfare Data from US Navy Contractor," Engadget, September 6, 2018, https://www.engadget.com/2018/06/09/chinese-hackers-stole-undersea-warfare-data-from-us-navy-contrac/.

10 Julian E. Barnes and Adam Goldman, "Russia Trying to Stoke U.S. Racial Tensions Before Election, Officials Say," New York Times, March 10, 2020, https://www.nytimes.com/2020/03/10/us/politics/russian-interference-race.html?fbclid=IwAR3Hh3xBe0zzhbg6AWewnl8xwSFJxF9nwUtCWBzp6_qlQ4jmkraqB6N1jgc

11 Alex Yablon, "Just How Many Guns Do Americans Actually Own?" Vice, June 25 2018 https://www.vice.com/en_us/article/bj3485/how-many-guns-are-there-in-america

12 Ylan Q. Mui, "The shocking number of Americans who can't cover a $400 expense," Washington Post, May 25, 2016 https://www.washingtonpost.com/news/wonk/wp/2016/05/25/the-shocking-number-of-americans-who-cant-cover-a-400-expense/

13 Richard Hason, "Trump's Jokes About Defying Election Results Could Create Chaos" Slate, Feb. 4, 2020, https://slate.com/news-and-politics/2020/02/trump-jokes-rigged-elections-chaos.html

14 United States of America v. Elena Alekseevna Khusyaynova, Eastern District of Virginia, Criminal Complaint Case #1:18-MJ-464, p. 16 https://assets.documentcloud.org/documents/5009827/Khusyaynova-Complaint.pdf

15 Cyberspace Solarium Commission Report, US Congress, March, 2020, 8

16 Ellen Nakashima and Seung Min Kim, "No evidence yet that Russia has taken steps to help any candidate in 2020, intelligence official tells Congress," Washington Post, March 10, 2020, https://www.washingtonpost.com/national-security/richard-grenell-trump-intelligence-election-security/2020/03/10/6504cc36-62d6-11ea-acca-80c22bbee96f_story.html

17 Robin Emmot, "Russia deploying coronavirus disinformation to sow panic in West, EU document says," Reuters, March 18, 2020, https://www.reuters.com/article/us-health-coronavirus-disinformation/russia-feeding-disinformation-about-coronavirus-to-sow-panic-in-west-eu-document-idUSKBN21518F?fbclid=IwAR2Hn1GWB1FrJ2mwY7zlhKHuHnZlWjCQ61FsBk8L_bVck9bR7UCcrqCiZ00

18 Jason Parham, "Targeting Black Americans, Russia's IRA Exploited Racial Wounds," Wired, December 17, 2018, https://www.wired.com/story/russia-ira-target-black-americans/

19 Terrell J. Starr, "How Russia Used Racism to Hack White Voters," The Root, September 8, 2017, https://www.theroot.com/how-russia-used-racism-to-hack-white-voters-1797582833.

20 Anthony Cuthbertson, "North Korean Hackers' Servers Seized in Thailand," The Independent, accessed November 28, 2019, https://www.independent.co.uk/life-style/gadgets-and-tech/news/north-korea-hackers-server-thailand-sony-pictures-cyber-attack-a8329586.html.

21 Daniel Brown, "Here's What China Plans to do with the J-31 Stealth Fighter — its Copy of the F-35," Business Insider, November 26, 2018, https://www.businessinsider.com/9-photos-of-j-31-chinas-copycat-version-of-the-f-35-stealth-fighter-2018-10

22 Thor Benson, "Trolls and bots are flooding social media with disinformation encouraging states to end quarantine," Business Insider, Apr 24, 2020, https://www.businessinsider.com/trolls-bots-flooding-social-media-with-anti-quarantine-disinformation-2020-4?fbclid=IwAR1IaZY_FU1zsNknZogR7PccqktcEpyl18sFmPOEt3YDjOKEP1AEU-G0RGwY

23 "2019 on track to being the "worst year on record" for breach activity." RiskBased Security, 2019 https://pages.riskbasedsecurity.com/2019-midyear-data-breach-quickview-report

24 Devon Milkovich, "15 Alarming Cybersecurity Facts and Stats," Cybint, September 23, 2019 https://www.cybintsolutions.com/cyber-security-facts-stats/

25 Allen Kim, "In the last 10 months, 140 local governments, police stations and hospitals have been held hostage by ransomware attacks," CNN , October 8, 2019 https://www.cnn.com/2019/10/08/business/ransomware-attacks-trnd/index.html

26 "Computers". Collier's Encyclopedia. Vol. 7, 1992: 114, 129.

27 "How Many Computers Are There in the World?" Reference.com, https://www.reference.com/world-view/many-computers-world-e2e980daa5e128d0

28 "Percentage of households in the United States with a computer at home from 1984 to 2016," Statista https://www.statista.com/statistics/214641/household-adoption-rate-of-computer-in-the-us-since-1997/

29 Aaron Smith, "Record shares of Americans now own smartphones, have home broadband," Pew Research, January 12, 2017, https://www.pewresearch.org/fact-tank/2017/01/12/evolution-of-technology/

30 Number of smartphone users worldwide from 2016 to 2021, Statista https://www.statista.com/statistics/330695/number-of-smartphone-users-worldwide/

31 Joseph Menn, "The Cult of the Dead Cow" PublicAffairs, 2019

32 Max Eddy, "SecurityWatch: Android vs. iOS , Which Is More Secure?" PCMag.com, April 24, 2019 https://www.pcmag.com/commentary/367918/securitywatch-android-vs-ios-which-is-more-secure

33 Rae Hodge, "iOS 13 vs. Android10: Which is more secure?" CNET , September 19, 2019 https://www.cnet.com/news/ios-13-vs-android-10-which-is-more-secure/

34 Craig Timberg, "A flaw in the design," The Washington Post, May 30, 2015 https://www.washingtonpost.com/sf/business/2015/05/30/net-of-insecurity-part-1/?utm_term=.88132ff9ec14

35 "Weapon Systems Cybersecurity:DOD Just Beginning to Grapple with Scale of Vulnerabilities" GAO-19-128: Published: Oct 9, 2018. Publicly Released: Oct 9, 2018, https://www.gao.gov/products/GAO-19-128

36 David Sanger, "The Perfect Weapon" 12

37 "FEDERAL CYBERSECURITY: AMERICA'S DATA AT RISK," United States Senate PERMANENT SUBCOMMITTEE ON INVESTIGATIONS Committee on Homeland Security and Governmental Affairs, 2019 https://www.portman.senate.gov/sites/default/files/2019-06/2019.06.25-PSI%20Report%20Final%20UPDATE.pdf

38 Matt Hamblen, "Clinton commits $1.46B to fight cyberterrorism," CNN , January 26, 1999 http://edition.cnn.com/TECH/computing/9901/26/clinton.idg/index.html

39 Maureen Webb, Coding Democracy: How Hackers are Disrupting Power, Surveillance and Authoritarianism, MIT Press, 2020

40 Steve McConnell, "Code Complete: A Practical Handbook of Software Construction, Second Edition," Microsoft Press, 2004

41 "Cyber-Attack Chain," BeyondTrust.com https://www.beyondtrust.com/resources/glossary/cyber-attack-chain

42 "Patch Management: Why it's Important for CyberSecurity," SwordShield.com, Oct 29, 2018, https://www.swordshield.com/blog/patch-management-important-cybersecurity/

43 Warwick Ashford, "Basic security could have prevented OPM breach," ComputerWeekly. com, September 7, 2016 https://www.computerweekly.com/news/450303826/Basic-security-could-have-prevented-OPM-breach-says-report

44 Verizon 2019 Data Breach Investigations Report, https://enterprise.verizon.com/resources/reports/dbir/

45 Symantec Internet Security Threat Report, Volume 23, March 2018 http://images.mktgassets.symantec.com/Web/Symantec/%7B3a70beb8-c55d-4516-98ed-1d0818a42661%7D_ISTR23_Main-FINAL-APR10.pdf?aid=elq

46 MIT Technology Review, "Ransomware may have cost the US more than $7.5 billion in 2019," Jan 2, 2019, https://www.technologyreview.com/f/615002/ransomware-may-have-cost-the-us-more-than-75-billion-in-2019/

47 Devon Milkovich, "15 Alarming Cybersecurity Facts and Stats," Cybint, September 23, 2019 https://www.cybintsolutions.com/cyber-security-facts-stats/

48 Jose Pagliery, "The inside story of the biggest hack in history," CNN Business, August 5, 2015 https://money.cnn.com/2015/08/05/technology/aramco-hack/index.html

49 Jordan Robertson and Michael Riley, "The Big Hack: How China Used a Tiny Chip to Infiltrate U.S. Companies," Bloomberg Businessweek, Oct 4, 2018, https://www.bloomberg.com/news/features/2018-10-04/the-big-hack-how-china-used-a-tiny-chip-to-infiltrate-america-s-top-companies

50 Steve Morgan, "Cybersecurity labor crunch to hit 3.5 million unfilled jobs by 2021," CSO, June 6, 2017, https://www.csoonline.com/article/3200024/cybersecurity-labor-crunch-to-hit-35-million-unfilled-jobs-by-2021.html

51 Jeffrey Anderson, "The Capital One Breach & "cloud_breach_s3" CloudGoat Scenario," Rhynosecuritylabs, 2019,https://rhinosecuritylabs.com/aws/capital-one-cloud_breach_s3-cloudgoat/

52 Mathew Ingram, "Legislation aimed at stopping deepfakes is a bad idea," Columbia Journalism Review, July 1, 2019, https://www.cjr.org/analysis/legislation-deepfakes.php

53 "Internet of Things - number of connected devices worldwide 2015-2025," Statista, https://www.statista.com/statistics/471264/iot-number-of-connected-devices-worldwide/

54 Bruce Schneier, "Using Hacked IoT Devices to Disrupt the Power Grid," Schneier on Security,September 11, 2018, https://www.schneier.com/blog/archives/2018/09/using_hacked_io.html

55 Josh Fruhlinger, "Top cybersecurity facts, figures and statistics for 2018," CSO, Oct 10, 2018, https://www.csoonline.com/article/3153707/top-cybersecurity-facts-figures-and-statistics.html

56 Symantec Internet Security Threat Report, Volume 24, February, 2019 https://www.symantec.com/security-center/threat-report

57 Dan Lohrmann, "2019: The Year Ransomware Targeted State & Local Governments," Government Technology, December 23, 2019, https://www.govtech.com/blogs/lohrmann-on-cybersecurity/2019-the-year-ransomware-targeted-state--local-governments.html

58 Nicole Karlis, "A huge security camera company just had a huge security breach," Salon, January 1, 2020, https://www.salon.com/2020/01/01/a-huge-security-camera-company-just-had-a-huge-security-breach/

59 Mathew J. Schwartz, "Salesforce Security Alert: API Error Exposed Marketing Data," Bank Info Security, August 3, 2018, https://www.bankinfosecurity.com/salesforce-security-alert-api-error-exposed-marketing-data-a-11278

60 Jeremy Haas & Ryan Bergquist, "Five Questions to Ask About Third-Party Vendors and Cybersecurity," SupplyChainBrain.com, November 19, 2019, https://www.supplychainbrain.com/blogs/1-think-tank/post/30489-the-company-you-keep-five-questions-to-ask-about-third-party-vendors-and-cybersecurity

61 George Friedman, "A U.S. Strategy Beyond the Cold War," RealClearWorld, May 16, 2016 https://www.realclearworld.com/articles/2016/05/16/a_us_strategy_beyond_the_cold_war_111854.html

62 Rob Marvin, "Tech Addiction By the Numbers: How Much Time We Spend Online," PC Magazine, June 11, 2018, https://www.pcmag.com/article/361587/tech-addiction-by-the-numbers-how-much-time-we-spend-online

63 Shadow War, 9

64 Stephanie Murray, "Putin: I wanted Trump to win the election," Politico, July 16, 2018, https://www.politico.com/story/2018/07/16/putin-trump-win-election-2016-722486

65 Maanvi Singh and Joan E Greve, "Republicans block election security bills after Mueller warns of Russian interference – as it happened," The Guardian, July 25, 2019, https://www.theguardian.com/us-news/live/2019/jul/25/trump-news-today-live-mueller-testimony-impeachment-calls-russia-latest-updates?page=with:block-5d39ff8c8f0845f89e3146c6

66 Philip Rucker, "Trump: Russian interference is 'all a big Dem HOAX!'", Washington Post, June 22, 2017, https://www.washingtonpost.com/news/post-politics/wp/2017/06/22/trump-russian-interference-is-all-a-big-dem-hoax/

67 Todd S. Purdum, "The Worst Russia Blunder in 70 Years," The Atlantic, July 20, 2018, https://www.theatlantic.com/politics/archive/2018/07/trump-russia-putin-helsinki/565652/

68 Greg Sargent, "Explosive new revelations just weakened Trump's impeachment defenses," Washington Post, Dec. 30, 2019, https://www.washingtonpost.com/opinions/2019/12/30/explosive-new-revelations-just-weakened-trumps-impeachment-defenses/?fbclid=IwAR1gSamLJV6y78vb3IQjCo5961MurdT2fUiOlrT7PcNSjHn5zW6P-I7UsDI

69 Shadow War, 11

70 Timothy D. Snyder, The Road to Unfreedom (London: Tim Duggan Books, 2018), 45.

71 Snyder – The Road to Unfreedom, 225

72 James Kirchick, "Russia's plot against the West. The Kremlin wants to destroy the trans-Atlantic alliance. Does Trump want to save it?" Politico, March 17, 2017, https://www.politico.eu/article/russia-plot-against-the-west-vladimir-putin-donald-trump-europe/

73 James Kirchick, "Russia's plot against the West. The Kremlin wants to destroy the trans-Atlantic alliance. Does Trump want to save it?" Politico, March 17, 2017, https://www.politico.eu/article/russia-plot-against-the-west-vladimir-putin-donald-trump-europe/

74 Snyder – The Road to Unfreedom, 226

75 Dustin Volz and Joseph Menn, "Russian hackers stole U.S. cyber secrets from NSA media reports," Reuters, October 5, 2017, https://www.reuters.com/article/us-usa-cyber-nsa-idUSKBN1CA2DO

76 Elbridge Colby and David Ochmanek, "How the United States Could Lose a Great-Power War," Foreign Policy, October 29, 2019, https://foreignpolicy.com/2019/10/29/united-states-china-russia-great-power-war/

77 Snyder – The Road to Unfreedom, 225

78 United States Senate, "(U)Report of the Select Committee on Intelligence United States Senate on Russian Active Measures Campaigns and Interference in the 2016 U.S. Election," 12 https://www.intelligence.senate.gov/sites/default/files/documents/Report_Volume2.pdf

79 Christian Caryl, "Donald Trump's talking points on Crimea are the same as Vladimir Putin's," Washington Post, July 3, 2018, https://www.washingtonpost.com/news/democracy-post/wp/2018/07/03/donald-trumps-talking-points-on-crimea-are-the-same-as-vladimir-putins/

80 Eric Geller, "Trump signs order setting stage to ban Huawei from U.S." Politico, May 15, 2019.https://www.politco.com/story/2019/05/15/trump-ban-huawei-us-1042046

81 Geoff Colvin, "Study: China Will Overtake the U.S. as World's Largest Economy Before 2030," Fortune, February 9, 2017, https://fortune.com/2017/02/09/study-china-will-overtake-the-u-s-as-worlds-largest-economy-before-2030/

82 Chinese Influence & American Interests: PROMOTING CONSTRUCTIVE VIGILANCE, Hoover Institution, 2018.https://www.hoover.org/sites/default/files/research/docs/ chineseinfluence_americaninterests_fullreport_web.pdf

83 China GDP Per Capita, Trading Economics, 2019,https://tradingeconomics.com/china/ gdp-per-capita

84 "How Come There Are So Many Billionaires in Communist China?" Bloomberg News, November 29, 2018, https://www.bloomberg.com/news/articles/2018-11-29/ why-communist-china-is-home-to-so-many-billionaires-quicktake

85 Rob Cooper, "Inside Apple's Chinese 'sweatshop' factory where workers are paid just £1.12 per hour to produce iPhones and iPads for the West," The Daily Mail, January 25, 2013 https:// www.dailymail.co.uk/news/article-2103798/Revealed-Inside-Apples-Chinese-sweatshop- factory-workers-paid-just-1-12-hour.html

86 Ryan, Pickrell, "China is building a powerful navy to take on the US in the Pacific — here's what its arsenal looks like," Business Insider, May 1, 2019, https://www.businessinsider.com. au/china-is-building-a-powerful-navy-to-take-on-the-us-in-the-pacific-2019-4

87 Ryan, Pickrell, "China is building a powerful navy to take on the US in the Pacific — here's what its arsenal looks like," Business Insider, May 1, 2019, https://www.businessinsider.com. au/china-is-building-a-powerful-navy-to-take-on-the-us-in-the-pacific-2019-4

88 Sasha Goldstein, "Chinese hackers stole F-35 fighter jet blueprints in Pentagon hack, Edward Snowden documents claim," NY Daily News, January 20, 2015, https://www.nydailynews.com/news/national/ snowden-chinese-hackers-stole-f-35-fighter-jet-blueprints-article-1.2084888

89 Robert D. Blackwill, "China's Strategy for Asia: Maximize Power, Replace America," The National Interest, May 26, 2016, https://nationalinterest.org/feature/ chinas-strategy-asia-maximize-power-replace-america-16359

90 Robert D. Blackwill, "China's Strategy for Asia: Maximize Power, Replace America," The National Interest, May 26, 2016, https://nationalinterest.org/feature/ chinas-strategy-asia-maximize-power-replace-america-16359

91 Eleanor Ross, "How and Why China is Building Islands in the South China Sea," Newsweek, March 29, 2017, https://www.newsweek.com/ china-south-china-sea-islands-build-military-territory-expand-575161

92 David Brunnstrom and Michael Martina, "Xi denies China turning artificial islands into military bases," Reuters, September 25, 2015, https://www.reuters.com/article/ us-usa-china-pacific-idUSKCN0RP1ZH20150925

93 Shadow war, 105

94 Stealth War, 12

95 Chinese Influence & American Interests: PROMOTING CONSTRUCTIVE VIGILANCE, Hoover Institution, 2018, 4, https://www.hoover.org/sites/default/files/research/docs/ chineseinfluence_americaninterests_fullreport_web.pdf

96 Most Important Problem, Gallup Poll, 2019, https://news.gallup.com/poll/1675/ Most-Important-Problem.aspx

97 Kim Parker, Anthony Cillufo and Renee Stepler, "6 facts about the U.S. military and its changing demographics," Pew Research Cetner, April 13, 2017, https://www.pewresearch.org/ fact-tank/2017/04/13/6-facts-about-the-u-s-military-and-its-changing-demographics/

98 Heather Long, "U.S. has lost 5 million manufacturing jobs since 2000," CNN Business, March 29, 2016, https://money.cnn.com/2016/03/29/news/economy/us-manufacturing-jobs/ index.html

99 Dina Smeltz and Lily Wojtowicz, "American Opinion on US-Russia Relations: From Bad to Worse," The Chicago Council on Global Affairs, August 2, 2017, https://www. thechicagocouncil.org/publication/american-opinion-us-russia-relations-bad-worse

100 Bill Whitaker, "How Russian Intelligence Officers Interfered in the 2016 Election," CBS News, November 24, 2019, https://www.cbsnews.com/news/russian-hackers-2016-election- democratic-congressional-campaign-committee-60-minutes-2019-11-24/?fbclid=IwAR1Be- bcUMUHa7SuOOlQ6mvuyJLdl3-84PyRgoYYP_jBL-9kY2taetOFDl4

101 "Yahoo Cyber Indictment Shows Kremlin, Hackers Working Hand-In-Hand," Deccan Chronicle, March 16, 2017, https://www.deccanchronicle.com/technology/ in-other-news/160317/yahoo-cyber-indictment-shows-kremlin-hackers-working-hand-in- hand.html

102 "Yahoo Cyber Indictment Shows Kremlin, Hackers Working Hand-In-Hand," Deccan Chronicle, March 16, 2017

103 Mark Galeotti, "Under Vladmir Putin, Gangsterism on the Streets Has Given Way to Kleptocracy in the State," The Guardian, March 23, 2018, https://www.theguardian.com/news/2018/mar/23/how-organised-crime-took-over-russia-vory-super-mafia

104 Mark Galeotti, "How the Kremlin Influences the West Using Russian Criminal Groups in Europe," Euromaidan Press, April 26, 2017, http://euromaidanpress.com/2017/04/26/how-the-kremlin-influences-the-west-using-russian-criminal-groups-in-europe/

105 Cory Bennett, "Kremlin's Ties to Russian Cyber Gangs Sow US Concerns," The Hill, October 11, 2015, https://thehill.com/policy/cybersecurity/256573-kremlins-ties-russian-cyber-gangs-sow-us-concerns

106 "Russian National Charged with Decade-Long Series of Hacking and Bank Fraud Offenses Resulting in Tens of Millions in Losses and Second Russian National Charged with Involvement in Deployment of 'Bugat' Malware," US Department of Justice, December 5, 2019, https://www.justice.gov/opa/pr/russian-national-charged-decade-long-series-hacking-and-bank-fraud-offenses-resulting-tens

107 Samuel Rubenfeld, "Russia-Based 'Evil Corp' Sanctioned by U.S., Top Members Charged after Probe with U.K.," Kharon Brief, December 05, 2019, https://brief.kharon.com/updates/russian-evil-corp-sanctioned-by-us-two-members-charged/

108 Rubenfeld, "Russia-Based 'Evil Corp' Sanctioned by U.S., Top Members Charged after Probe with U.K.,"

109 Bojana Dobran, "27 Terrifying Ransomware Statistics & Facts You Need to Read," PhoenixNAP, January 31, 2019, https://phoenixnap.com/blog/ransomware-statistics-facts

110 Frances Robles, "A City Paid a Hefty Ransom to Hackers. But Its Pains Are Far from over," New York Times, July 7, 2019, https://www.nytimes.com/2019/07/07/us/florida-ransom-hack.html

111 Danny Palmer, "Ransomware: Cyber- Insurance Payouts are Adding to the Problem, Warn Security Experts," ZDNet, September 17, 2019, https://www.zdnet.com/article/ransomware-cyber-insurance-payouts-are-adding-to-the-problem-warn-security-experts/

112 "State of Ransomware in the U.S.: 2019 Report for Q1 to Q3," Emsisoft, October 1, 2019, https://blog.emsisoft.com/en/34193/state-of-ransomware-in-the-u-s-2019-report-for-q1-to-q3/

113 Manny Fernandez, David E. Sanger, and Marina Trahan Martinez, "Ransomware Attacks Are Testing Resolve of Cities Across America," New York Times, August 22, 2019, https://www.nytimes.com/2019/08/22/us/ransomware-attacks-hacking.html

114 Benjamin Freed, "Recent Ransomware Surge Linked to Russian Criminal Group," Statescoop, September 3, 2019, https://statescoop.com/recent-ransomware-surge-russian-criminal-group/

115 Naveen Goud, "Maze Ransomware was behind the Cyberattack of Pensacola Florida," Cybersecurity Insiders, December 7, 2019, https://www.cybersecurity-insiders.com/maze-ransomware-was-behind-the-cyber-attack-of-pensacola-florida/

116 Catalin Cimpanu, "New Orleans hit by ransomware, city employees told to turn off computers," December 13, 2019 Security, https://www.zdnet.com/article/new-orleans-hit-by-ransomware-city-employees-told-to-turn-off-computers/

117 "New Orleans Declares State of Emergency After Ransomware Attack," CISO Magazine, December 17, 2019 https://www.cisomag.com/new-orleans-declares-state-of-emergency-after-ransomware-attack/

118 David Voreacos, Katherine Chiglinsky, and Riley Griffin, "Merck Cyberattack's $1.3 Billion Question: Was It an Act of War?" Claims Journal, December 3, 2019, https://www.claimsjournal.com/news/national/2019/12/03/294354.htm

119 Olivia Solon, "'Petya' Ransomware Attack: What Is It and How Can It Be Stopped?," The Guardian, June 28, 2017, https://www.theguardian.com/technology/2017/jun/27/petya-ransomware-cyber-attack-who-what-why-how

120 Iain Thomson, "Everything You Need to Know about the Petya, er, NotPetya Nasty Trashing PCs Worldwide," The Register, June 28, 2017, https://www.theregister.co.uk/2017/06/28/petya_notpetya_ransomware/

121 "Study: Ransomware, Data Breaches at Hospitals tied to Uptick in Fatal Heart Attacks," Krebs on Security, November 7, 2019, https://krebsonsecurity.com/2019/11/study-ransomware-data-breaches-at-hospitals-tied-to-uptick-in-fatal-heart-attacks/

122 Kaja Whitehouse, "Russian hacker accused of mass data breach extradited to US," New York Post, September 7, 2018, https://nypost.com/2018/09/07/russian-hacker-accused-of-mass-data-breach-extradited-to-us/

123 Warwick Ashford, "US Sentences Two Russians for Huge Data Breaches," Computer Weekly, February 16, 2018, https://www.computerweekly.com/news/252435193/US-sentences-two-Russians-for-huge-data-breaches

124 "Significant Cyber Incidents," Center for Strategic and International Studies, accessed December 11, 2019, https://www.csis.org/programs/technology-policy-program/significant-cyber-incidents

125 Max Fischer, "In Case You Weren't Clear on Russia Today's Relationship to Moscow, Putin Clears it up," Washington Post, June 13, 2013, https://www.washingtonpost.com/news/worldviews/wp/2013/06/13/in-case-you-werent-clear-on-russia-todays-relationship-to-moscow-putin-clears-it-up/

126 Cliff Kincaid, "Why Won't Putin Help Middle East Christians?," Accuracy in Media, August 22, 2014, https://www.aim.org/aim-column/why-wont-putin-help-middle-east-christians/

127 Sonia Scherr, "Russian TV Channel Pushes 'Patriot' Conspiracy Theories," Southern Poverty Law Center, August 01, 2010, https://www.splcenter.org/fighting-hate/intelligence-report/2010/russian-tv-channel-pushes-patriot-conspiracy-theories

128 Samantha Bradshaw and Philip N. Howard, The Global Disinformation Order 2019 Global Inventory of Organised Social Media Manipulation (Oxford: University of Oxford, 2019), https://comprop.oii.ox.ac.uk/wp-content/uploads/sites/93/2019/09/CyberTroop-Report19.pdf

129 Agence France-Presse, "Facebook Removed 5.4 Billion Fake Accounts this Year," South China Morning Post, November 14, 2019, https://www.scmp.com/news/world/united-states-canada/article/3037641/facebook-removed-54-billion-fake-accounts-year

130 Center for Strategic and International Studies, "Significant Cyber Incidents."

131 David E. Sanger, "Russian Hackers Appear to Shift Focus to U.S. Power Grid," New York Times, July 27, 2018 https://www.nytimes.com/2018/07/27/us/politics/russian-hackers-electric-grid-elections-.html

132 Kim Zetter, 'Inside the Cunning, Unprecedented Hack of Ukraine's Power Grid," Wired, March 3, 2016 https://www.wired.com/2016/03/inside-cunning-unprecedented-hack-ukraines-power-grid/

133 Harry Morgan, "Utilities Must Stop Cutting Corners as Cyberthreats Rise," Rethink, November 14, 2019https://rethinkresearch.biz/articles/utilities-must-stop-cutting-corners-as-cyberthreats-rise-2/

134 DH Kass, "LookBack Malware Attacked Energy Utilities Across 18 States," MSSP Alert, January 2, 2020 https://www.msspalert.com/cybersecurity-markets/verticals/lookback-malware-targeted-energy-utilities/

135 Maggie Miller, "Federal council to Trump: Cyber threats pose 'existential threat' to the nation," The Hill, December 9, 2019 https://thehill.com/policy/cybersecurity/473682-federal-council-to-trump-cyber-threats-pose-existential-threat-to-the?fbclid=IwAR04Gfg10cJfp-7a1JP6ygph8l3m18zgpCw9cF6FI9DFEE3FtEjOx4sr4DY

136 Bobby Allyn, "Researchers: Nearly Half Of Accounts Tweeting About Coronavirus Are Likely Bots," NPR.org, May 20, 2020, https://www.npr.org/sections/coronavirus-live-updates/2020/05/20/859814085/researchers-nearly-half-of-accounts-tweeting-about-coronavirus-are-likely-bots?fbclid=IwAR1ROpJ5c8bR5xMlLWj5PEvmGzhxqENeNGvnDdcW3MAxkaef4qo7ZVmMItU

137 Edward Wong, Matthew Rosenberg and Julian E. Barnes, "Chinese Agents Helped Spread Messages That Sowed Virus Panic in U.S., Officials Say," New York Times, April 22, 2020, https://www.nytimes.com/2020/04/22/us/politics/coronavirus-china-disinformation.html

138 Ali Breland, "Thousands attended protest organized by Russians on Facebook," The Hill, Oct. 31, 2017, https://thehill.com/policy/technology/358025-thousands-attended-protest-organized-by-russians-on-facebook

139 Claire Allbright, "A Russian Facebook Page Organized a Protest in Texas." Texas Tribune, November 1, 2017 https://www.texastribune.org/2017/11/01/russian-facebook-page-organized-protest-texas-different-russian-page-l/

140 Julian E. Barnes and Adam Goldman, "Russia Trying to Stoke U.S. Racial Tensions Before Election, Officials Say," New York Times, March 10, 2020 https://www.nytimes.

com/2020/03/10/us/politics/russian-interference-race.html?fbclid=IwAR34MeZ5e5yEaA0Jurt
rqpubdu0Lc5ELs8MJwjwY7iTI5-PpqJ1ezZ-wRm8

141 Andrew Higgins, "Russia's Dark Arts," New York Times, May 30, 2016, https://www.nytimes.
com/series/russias-dark-arts?module=inline

142 Michael Schwirtz, "Top Secret Russian Unit Seeks to Destabilize Europe, Security Officials
Say," New York Times, October 8, 2019, https://www.nytimes.com/2019/10/08/world/
europe/unit-29155-russia-gru.html?fbclid=IwAR2ZwGl1IKMgHb1pO0VFYhXZnRl46OjIdxj
Lm3gPYUbDFsgzvbM8y3FGSc0

143 United States Senate, "(U)Report of the Select Committee on Intelligence United States Senate
on Russian Active Measures Campaigns and Interference in the 2016 U.S. Election," 11.

144 Snyder, The Road to Unfreedom, 86.

145 M.B. Pell and Echo Wang, "U.S. Navy bans TikTok from government-issued mobile devices,"
Reuters, December 20, 2019, https://www.reuters.com/article/us-usa-tiktok-navy/u-s-navy-
bans-tiktok-from-government-issued-mobile-devices-idUSKBN1YO2HU

146 Andrew Hutchinson, "TikTok Banned for US Military Personnel as Questions Continue to
Swirl Around the App," Social Media Today, Jan 2, 2020, https://www.socialmediatoday.com/
news/tiktok-banned-for-us-military-personnel-as-questions-continue-to-swirl-arou/569652/

147 Chris Mills Rodrigo, "Schumer, Cotton request TikTok security
assessment," The Hill, October 24, 2019, https://thehill.com/policy/
technology/467280-schumer-cotton-request-tiktok-security-assessment

148 Anna Wilde Mathews, "Anthem: Hacked Database Included 78.8 Million
People," Wall Street Journal, Feb. 24, 2015, https://www.wsj.com/articles/
anthem-hacked-database-included-78-8-million-people-1424807364

149 "Anthem to Pay Record $115M to Settle Lawsuits Over Data Breach," NBC
News/Reuters, June 23, 2017, https://www.nbcnews.com/news/us-news/
anthem-pay-record-115m-settle-lawsuits-over-data-breach-n776246

150 Christopher Porter and Brian Finch, "What Does Beijing Want With Your Medical
Records?" Wall Street Journal, June 20, 2019, https://www.wsj.com/articles/
what-does-beijing-want-with-your-medical-records-11561069899

151 Kate Fazzini, "The great Equifax mystery: 17 months later, the stolen data has never been
found, and experts are starting to suspect a spy scheme," CNBC, Feb 13, 2019, https://www.
cnbc.com/2019/02/13/equifax-mystery-where-is-the-data.html

152 US Department of Justice, "Chinese Military Personnel Charged with Computer Fraud,
Economic Espionage and Wire Fraud for Hacking into Credit Reporting Agency Equifax,"
February 10, 2020, https://www.justice.gov/opa/pr/chinese-military-personnel-charged-
computer-fraud-economic-espionage-and-wire-fraud-hacking

153 Matt Apuzzo and Michael S. Schmidt, "Secret Back Door in Some U.S. Phones Sent Data to
China, Analysts Say," New York Times, Nov. 15, 2016, https://www.nytimes.com/2016/11/16/
us/politics/china-phones-software-security.html

154 Jai Vijayan, "Chinese Cyber Espionage Group Steals SMS Messages via Telco Networks,"
Dark Reading, October 31, 2019, https://www.darkreading.com/attacks-breaches/
chinese-cyber-espionage-group-steals-sms-messages-via-telco-networks/d/d-id/1336235

155 "Report: Nearly 40% of Security Cameras May Be Vulnerable to Cyber-Attacks," Security
Sales & Integration, December 09, 2019, https://www.securitysales.com/research/
report-security-cameras-vulnerable-cyber/

156 "Smart Spies: Alexa and Google Home expose users to vishing and eavesdropping," Security
Research Labs, December 17, 2019, https://srlabs.de/bites/smart-spies/

157 Zack Whittaker, "Now even the FBI is warning about your smart TV's security," TechCrunch,
December 1, 2019, https://techcrunch.com/2019/12/01/fbi-smart-tv-security/

158 "Google cyber attack hit password system" NY Times, Reuters, April 19, 2010

159 Jacobs, Andrew; Helft, Miguel (January 12, 2010). "Google, Citing Attack, Threatens to Exit
China". The New York Times.

160 Josh Rogin, "NSA Chief: Cybercrime constitutes the 'greatest transfer of wealth
in history'" Foreign Policy, July 9, 2012, https://foreignpolicy.com/2012/07/09/
nsa-chief-cybercrime-constitutes-the-greatest-transfer-of-wealth-in-history/

161 James Cook, "FBI Director: China Has Hacked Every Big US Company,"
Business Insider, Oct 6, 2014, https://www.businessinsider.com/
fbi-director-china-has-hacked-every-big-us-company-2014-10

162 Stealth war, 146

163 Stealth War, 147

164 Finkle, J., Menn, J., Viswanatha, J. U.S. accuses China of cyber spying on American companies. Reuters, Mon May 19, 2014

165 FBI warns U.S. businesses of cyber attacks, blames BeijingOctober 16, 2014 Reuters

166 Teresa Welsh, "Obama, Xi Reach Agreement to End Cyberattacks," US News, Sept. 25, 2015, https://www.usnews.com/news/articles/2015/09/25/president-obama-chinese-president-xi-jingping-announce-agreement-to-stop-hacking

167 Jethro Mullen, "Chinese hackers are ramping up attacks on US companies," CNN Business, February 20, 2019, https://www.cnn.com/2019/02/20/tech/crowdstrike-china-hackers-us/index.html

168 Lesley Stahl, "The Great Brain Robbery: Economic espionage sponsored by the Chinese government is costing U.S. corporations hundreds of billions of dollars and more than two million jobs, " 60 Minutes, January 17, 2016, https://www.cbsnews.com/news/60-minutes-great-brain-robbery-china-cyber-espionage/

169 Rob Barry and Dustin Volz, "Ghosts in the Clouds: Inside China's Major Corporate Hack," Wall Street Journal, Dec. 30, 2019, https://www.wsj.com/articles/ghosts-in-the-clouds-inside-chinas-major-corporate-hack-11577729061

170 Jon Swaine and David Smith, "Suspected bombs sent to prominent Trump critics 'an act of terror'" The Guardian, October 24, 2018, https://www.theguardian.com/us-news/2018/oct/24/clinton-bomb-reports-delivery-home-new-york-hillary-bill-latest

171 Sam Sanders, "Massive Data Breach Puts 4 Million Federal Employees' Records At Risk," NPR, June 4, 2015, https://www.npr.org/sections/thetwo-way/2015/06/04/412086068/massive-data-breach-puts-4-million-federal-employees-records-at-risk

172 Andrea Peterson, "OPM says 5.6 million fingerprints stolen in biggest cyber attack in US history," Independent, September 24, 2015, https://www.independent.co.uk/news/world/americas/opm-says-56-million-fingerprints-stolen-in-biggest-cyber-attack-in-us-history-10515256.html

173 Ellen Nakashima, "Hacks of OPM databases compromised 22.1 million people, federal authorities say," Washington Post, July 9, 2015, https://www.washingtonpost.com/news/federal-eye/wp/2015/07/09/hack-of-security-clearance-system-affected-21-5-million-people-federal-authorities-say/

174 "Anomali Threat Research Team Identifies Widespread Credential Theft Campaign Aimed at U.S. and International Government Agency Procurement Services," December 12, 2019, https://finance.yahoo.com/news/anomali-threat-research-team-identifies-100010254.html

175 Claburn, Thomas. "China Cyber Espionage Threatens U.S., Report Says". InformationWeek. February 27, 2010

176 Claburn, Thomas. "China Cyber Espionage Threatens U.S., Report Says". InformationWeek February 27, 2010

177 Cha, Ariana Eunjung and Ellen Nakashima, "Google China cyberattack part of vast espionage campaign, experts say," The Washington Post, January 14, 2010

178 Ros Krasny, "Chinese hacked U.S. military contractors, Senate panel finds," Reuters, September 17, 2014, https://www.reuters.com/article/us-usa-military-cyberspying/chinese-hackers-breach-u-s-military-contractors-senate-probe-idUSKBN0HC1TA20140917

179 Jeremy Kirk, "US Army Nixes Use of DJI Drones Over Cybersecurity Concerns," Data Breach Today, August 7, 2017, https://www.databreachtoday.com/us-army-nixes-use-dji-drones-over-cybersecurity-concerns-a-10170

180 Department of the Navy letter: "Operating Risks with Regards to DJI Family of Products," May 24, 2017, https://nsarchive2.gwu.edu/dc.html?doc=6574684-National-Security-Archive-Department-of-the-Navy

181 Jed Pressgrove, "Bill Could Block Federal Money for Drones Made in China," Government Technology, May 27, 2020 https://www.govtech.com/policy/Bill-Could-Block-Federal-Money-for-Drones-Made-in-China.html

182 Ellen Nakashima and Paul Sonne, "China hacked a Navy contractor and secured a trove of highly sensitive data on submarine warfare," Washington Post, June 8, 2018, https://www.washingtonpost.com/world/national-security/china-hacked-a-navy-contractor-and-secured-a-trove-of-highly-sensitive-data-on-submarine-warfare/2018/06/08/6cc396fa-68e6-11e8-bea7-c8eb28bc52b1_story.html

183 Gordon Lubold and Dustin Volz, "Navy, Industry Partners Are 'Under Cyber Siege' by Chinese Hackers, Review Asserts Hacking threatens U.S.'s standing as world's leading military power, study says," Wall Street Journal March 12, 2019, https://www.wsj.com/articles/navy-industry-partners-are-under-cyber-siege-review-asserts-11552415553

184 Nicole Hong, "A Military Camera Said 'Made in U.S.A.' The Screen Was in Chinese." New York Times, Nov. 7, 2019, https://www.nytimes.com/2019/11/07/nyregion/aventura-china-cameras.html?action=click&module=Top%20Stories&pgtype=Homepage

185 Cooper Quintinmarch, "Are Your Devices Hardwired For Betrayal?" Electronic Frontier Foundation, March 2, 2015, https://www.eff.org/deeplinks/2015/03/hardwired-for-betrayal

186 Jordan Robertson and Michael Riley, "The Big Hack: How China Used a Tiny Chip to Infiltrate U.S. Companies" Bloomberg Cybersecurity, October 4, 2018, https://www.bloomberg.com/news/features/2018-10-04/the-big-hack-how-china-used-a-tiny-chip-to-infiltrate-america-s-top-companies

187 Andy Greenberg, "Planting Tiny Spy Chips in Hardware Can Cost as Little as $200" Wired, October 10, 2018, https://www.wired.com/story/plant-spy-chips-hardware-supermicro-cheap-proof-of-concept/

188 Jonathan Hillman, "Pretending all Chinese companies are evil schemers will only hurt the U.S. economy" Washington Post, November 8, 2019, https://www.washingtonpost.com/outlook/pretending-all-chinese-companies-are-evil-schemers-will-only-hurt-the-us-economy/2019/11/08/b0d98798-00dc-11ea-9518-1e76abc088b6_story.html

189 Bethany Biron, "The last decade was devastating for the retail industry. Here's how the retail apocalypse played out." Business Insider, Dec 23, 2019, https://www.businessinsider.com/retail-apocalypse-last-decade-timeline-2019-12

190 Susan B. Cassidy, "FISMA Updated and Modernized," Inside Government Contracts (Covington & Burling LLP), December 19, 2014, https://www.insidegovernmentcontracts.com/2014/12/fisma-updated-and-modernized/

191 Kyle Mizokami, "Russia vs. America: Which Army Would Win a War?" The National Interest, May 29, 2018, https://nationalinterest.org/blog/the-buzz/russia-vs-america-which-army-would-win-war-26008

192 Arthur Allen, "POLITICO-Harvard poll: Americans worried about data hacks, want higher taxes on e-cigs," Politico, August 26, 2019, https://www.politico.com/story/2019/08/26/politico-harvard-poll-data-hacks-e-cigs-1677630

193 Gordon Corera, "How France's TV5 was almost destroyed by 'Russian hackers'," BBC News 10 October 2016, https://www.bbc.com/news/technology-37590375

194 "GUESS WHAT REQUIRES 150 MILLION LINES OF CODE...." eit Digital, January 13, 2016, https://www.eitdigital.eu/newsroom/blog/article/guess-what-requires-150-million-lines-of-code/

195 "Connected Cars – Statistics and Facts," Statista, August 27, 2018, https://www.statista.com/topics/1918/connected-cars/

196 "10 IoT Security Incidents That Make You Feel Less Secure," CISO Mag, January 10, 2020, https://www.cisomag.com/10-iot-security-incidents-that-make-you-feel-less-secure/

197 Susan N. Houseman, "Is Automation Really to Blame for Lost Manufacturing Jobs?" Foreign Affairs, September 7, 2018, https://www.foreignaffairs.com/articles/2018-09-07/automation-really-blame-lost-manufacturing-jobs

198 Carrie Dann, "A deep and boiling anger': NBC/WSJ poll finds a pessimistic America despite current economic satisfaction," NBC News, August 25, 2019, https://www.nbcnews.com/politics/meet-the-press/deep-boiling-anger-nbc-wsj-poll-finds-pessimistic-america-despite-n1045916

199 Nicholas Kristof and Sheryl WuDunn, "Who Killed the Knapp Family?" New York Times, Jan. 9, 2020, https://www.nytimes.com/2020/01/09/opinion/sunday/deaths-despair-poverty.html?smid=nytcore-ios-share&fbclid=IwAR1-iDhDVZ4jj14CyTXcWJm8da6cX22lVEZgEObg4QNLDBrNw_EedV1m7PI

200 "US drugs bust uncovers enough Chinese fentanyl 'to kill 14 million people'" Associated Press, August 30, 2019, https://www.scmp.com/news/world/united-states-canada/article/3024993/us-drugs-bust-uncovers-enough-chinese-fentanyl-kill

201 Nicholas Carr, The Shallows, 3

202 "Evening Network News Ratings," Pew Research Center, March 13, 2006, https://www.journalism.org/numbers/network-evening-news-ratings/

203 J. Katz, "Here Are Evening News Ratings for the 2018-2019 Season and 3rd Quarter of 2019," Adweek, Sep. 24, 2019, https://www.adweek.com/tvnewser/here-are-the-evening-news-ratings-for-2018-2019-and-for-the-3rd-quarter-of-2019/415283/

204 "Leading cable news networks in the United States in October 2019, by number of primetime viewers," Statista, https://www.statista.com/statistics/373814/cable-news-network-viewership-usa/

205 Paul Bond, "Leslie Moonves on Donald Trump: 'It May Not Be Good for America, but It's Damn Good for CBS'" Hollywood Reporter, February 29, 2016, https://www.hollywoodreporter.com/news/leslie-moonves-donald-trump-may-871464

206 New York Times Display Advertising Report, SimilarWeb, https://www.similarweb.com/website/nytimes.com#display

207 Drew Harwell, "Faked Pelosi videos, slowed to make her appear drunk, spread across social media," Washington Post May 24, 2019, https://www.washingtonpost.com/technology/2019/05/23/faked-pelosi-videos-slowed-make-her-appear-drunk-spread-across-social-media/

208 "Leading cable news networks in the United States in October 2019, by number of primetime viewers," Statista, https://www.statista.com/statistics/373814/cable-news-network-viewership-usa/

209 Elisa Shearer and Katerina Eva Matsa, "News Use Across Social Media Platforms 2018: Most Americans continue to get news on social media, even though many have concerns about its accuracy," Pew Research Center, September 10, 2018, https://www.journalism.org/2018/09/10/news-use-across-social-media-platforms-2018/

210 "Donald Trump on social media," Wikipedia https://en.wikipedia.org/wiki/Donald_Trump_on_social_media

211 Tom Huddleston Jr., "This is how much it costs to air a commercial during the 2019 Super Bowl," CNBC, Jan 30 2019, https://www.cnbc.com/2019/01/30/how-much-it-costs-to-air-a-commercial-during-super-bowl-liii.html

212 "Donald Trump on social media," Wikipedia https://en.wikipedia.org/wiki/Donald_Trump_on_social_media

213 Mike Ausford, "Is Online Advertising Expensive?" Topdraw.com, March 1, 2019, https://www.topdraw.com/blog/is-online-advertising-expensive/

214 Kevin Roose, "On Gab, an Extremist-Friendly Site, Pittsburgh Shooting Suspect Aired His Hatred in Full," New York Times, Oct. 28, 2018, https://www.nytimes.com/2018/10/28/us/gab-robert-bowers-pittsburgh-synagogue-shootings.html

215 "Propaganda-spewing Russian trolls act differently online from regular people," The Conversation, September 5, 2018, http://theconversation.com/propaganda-spewing-russian-trolls-act-differently-online-from-regular-people-100855

216 Sean Illing "'Flood the zone with shit': How misinformation overwhelmed our democracy," Vox.com, January 18, 2020, https://www.vox.com/policy-and-politics/2020/1/16/20991816/impeachment-trial-trump-bannon-misinformation

217 Megan McArdle, "We finally know for sure that lies spread faster than the truth. This might be why." Washington Post, March 14, 2018, https://www.washingtonpost.com/opinions/we-finally-know-for-sure-that-lies-spread-faster-than-the-truth-this-might-be-why/2018/03/14/92ab1aae-27a6-11e8-bc72-077aa4dab9ef_story.html

218 Maggie Fox, "Fake News: Lies spread faster on social media than truth does," NBC News, March 8, 2018, https://www.nbcnews.com/health/health-news/fake-news-lies-spread-faster-social-media-truth-does-n854896

219 Hannah Ritchie, "Read all about it: The biggest fake news stories of 2016," CNBC, Dec, 30, 2016, https://www.cnbc.com/2016/12/30/read-all-about-it-the-biggest-fake-news-stories-of-2016.html

220 James Leggate, "Facebook users believe more than half of fake news is true, study finds," FOX Business, November 8, 2019, https://www.foxbusiness.com/technology/facebook-users-believe-fake-news-study

221 Todd Spangler, "Are Americans Addicted to Smartphones? U.S. Consumers Check Their Phones 52 Times Daily, Study Finds," Variety, November 14, 2018, https://variety.com/2018/digital/news/smartphone-addiction-study-check-phones-52-times-daily-1203028454/

222 By Patrick Nelson, "We touch our phones 2,617 times a day, says study," Network World, July 7, 2016, https://www.networkworld.com/article/3092446/we-touch-our-phones-2617-times-a-day-says-study.html

223 Anderson Cooper, "What is "brain hacking"? Tech insiders on why you should care," 60 Minutes, April 9, 2017, https://www.cbsnews.com/news/brain-hacking-tech-insiders-60-minutes/

224 Chauncey DeVega, "Cult expert Steven Hassan: Trump's "mind control cult" now faces an existential crisis," Salon, April 7, 2020, https://www.salon.com/2020/04/07/cult-expert-steven-hassan-trumps-mind-control-cult-now-faces-an-existential-crisis/

225 Alexandra Levin and Zach Montellaro, "Facebook sticking with policies on politicians' lies and voter targeting," Politico, January 9, 2020 https://www.politico.com/news/2020/01/09/facebook-sticking-with-policies-on-politicians-lies-and-voter-targeting-096597?fbclid=IwAR2uHMsl1_npJTZp6Xyjeo63XIWLeIuGRrSIOEa6g-B8mhcZdC87FbNYSN0

226 O'Connor and Weatherall The Misinformation Age: How False Beliefs Spread, Yale University Press February 18, 2020

227 Dahlia Lithwick, "What to Make of the Jarring Return to News in 2020," Slate, January 9, 2020, https://slate.com/news-and-politics/2020/01/return-to-news-2020-trump-iran-war-real.html?fbclid=IwAR3i3ub9YMvb0mAiebrrJPDtpTWA4lvL9lWEP3BOI0GB-f_96Qi5GELkO88

228 Snyder – Road to Unfreedom p 105

229 Alex Henderson, "Here are 7 public figures who have received death threats for criticizing Donald Trump," Alternet, August 6, 2018, https://www.rawstory.com/2018/08/7-public-figures-received-death-threats-criticizing-trump/

230 Daniel Berti, "Carter says he'll skip Monday's General Assembly session following death threats," Prince William Times, January 17, 2020, https://www.princewilliamtimes.com/news/carter-says-he-ll-skip-monday-s-general-assembly-session/article_a02949f0-3936-11ea-8828-ef83c88d22c7.html

231 Garrett Epps, "Guns Are No Mere Symbol," The Atlantic, January 21, 2020, https://www.theatlantic.com/ideas/archive/2020/01/guns-are-no-mere-symbol/605239/

232 Sebastian Murdock and Andy Campbell, "The Colossal Gun Rally In Richmond Was Co-Opted By Extremists," Huffington Post, January 23, 2020, https://www.huffpost.com/entry/richmond-gun-rally-extremists_n_5e2663a1c5b6321176184b6f

233 Ted Dunlap, "Modern slang definition: The Boogaloo," Bitterroot Bugle, December 1, 2019, https://www.bitterrootbugle.com/2019/12/01/modern-slang-definition-the-boogaloo/

234 Craig Mauger, "Whitmer denounces death threats, wants 450,000 COVID-19 tests in May," The Detroit News, May 11, 2020 https://www.detroitnews.com/story/news/local/michigan/2020/05/11/whitmer-updates-state-covid-19-response/3110530001/

235 "Ransomware attack takes US maritime base offline," BBC News, January 2, 2020, https://www.bbc.com/news/technology-50972890

236 David Sanger, The Perfect weapon, p 19

237 Mark Pomerleau, "How hackers breached an Air Force system," Fifth Domain, December 12, 2019, https://www.fifthdomain.com/dod/air-force/2019/12/12/how-hackers-breached-an-air-force-system/

238 Mark Pomerleau, "How hackers breached an Air Force system," Fifth Domain, December 12, 2019, https://www.fifthdomain.com/dod/air-force/2019/12/12/how-hackers-breached-an-air-force-system/

239 Shawn Campbell, "Military Security in the Age of the Internet of Things," ACFEA Signal, February 1, 2016, https://www.afcea.org/content/?q=Article-military-security-age-internet-things

240 Paul Szoldra, "How the US military is beating hackers at their own game," Business Insider, May 24, 2016, https://www.businessinsider.com/us-military-cyberwar-2016-5

241 Elbridge Colby and David Ochmanek, "How the United States Could Lose a Great-Power War," Foreign Policy, October 29, 2019, https://foreignpolicy.com/2019/10/29/united-states-china-russia-great-power-war/

242 Department of Defense, Defense Science Board "Resilient Military Systems and the Advanced Cyber Threat" 2013, p 5

243 "Officials: Hack exposed U.S. military and intel data," CBS News, June 12, 2015, https://www.cbsnews.com/news/officials-second-u-s-government-hack-exposed-military-and-intel-data/

244 T. Christian Miller, Megan Rose and Robert Faturechi, "Fight The Ship: Death and valor on a warship doomed by its own Navy." ProPublica, February 6, 2019, https://features.propublica.org/navy-accidents/uss-fitzgerald-destroyer-crash-crystal/

245 "Navy Expands Cyber Warrant Program," Navy.mil, June 4, 2018, https://www.navy.mil/submit/display.asp?story_id=105858

246 Joanna Crews, "US Navy Launches Inspection Program for Cyber Operations Preparedness," GovConDaily, June 6, 2018, https://www.executivegov.com/2018/06/us-navy-launches-inspection-program-for-cyber-operations-preparedness/

247 Alex Hern, "Fitness tracking app Strava gives away location of secret US army bases," The Guardian, January 28, 2018, https://www.theguardian.com/world/2018/jan/28/fitness-tracking-app-gives-away-location-of-secret-us-army-bases

248 Hugh Taylor, "The Polar Fitness Tracker Episode and the Frustrating Pace of Military Cyber Policy Change," Journal of Cyber Policy, July 11, 2018, https://journalofcyberpolicy.com/2018/07/11/polar-fitness-tracker-episode-frustrating-pace-military-cyber-policy-change/

249 Kevin Breuningerm, "'Frankly, the United States is under attack': DNI Coats sounds alarm over cyberthreats from Russia," CNBC, February 13, 2018, https://www.cnbc.com/2018/02/13/the-united-states-is-under-attack-coats-warns-of-cyber-threats.html

250 Isobel Asher Hamilton, "A fitness app exposed sensitive location details for thousands of users including soldiers and secret agents," Business Insider, Jul 9, 2018, https://www.businessinsider.com/polar-exercise-fitness-app-exposed-soldiers-spies-location-details-2018-7/

251 Lily Hay Newman, "Russia Takes a Big Step Toward Internet Isolation," Wired, January 5, 2020, https://www.wired.com/story/russia-internet-control-disconnect-censorship/

252 Mark Pomerleau, "What the budget request explains about Cyber Command's goals," Fifth Domain, February 20, 2018, https://www.fifthdomain.com/dod/2018/02/20/what-the-budget-request-explains-about-cyber-commands-goals/

253 "31% Think U.S. Civil War Likely Soon," Rasmussen Reports, June 27, 2018, http://www.rasmussenreports.com/public_content/politics/general_politics/june_2018/31_think_u_s_civil_war_likely_soon

254 Tess Owen, "Far-Right Extremists Are Hoping to Turn the George Floyd Protests Into a New Civil War," Vice, May 29 2020 https://www.vice.com/en_us/article/pkyb9b/far-right-extremists-are-hoping-to-turn-the-george-floyd-protests-into-a-new-civil-war?fbclid=IwAR3nCV-TqUw25NXCEhQbcZn-_jGjK9lBpE_CZ1VULcoEf2KOiYSMAl4dwkE

255 Michaek Kunzelman, "Documents: Extremist group wanted rally to start civil war," AP, January 21, 2020, https://apnews.com/e5d17a8735678aa604a22f011c2685db?fbclid=IwAR1h mYUmTqdWKHtbnTxJqzMjZQwWZ6fc5yms94_BmSCV3z-4uDMll-SGSpw

256 Daniel De Simone, Andrei Soshnikov & Ali Winston, "Neo-Nazi Rinaldo Nazzaro running US militant group The Base from Russia," BBC News, January 24, 2020, https://www.bbc.com/news/world-51236915?fbclid=IwAR3qHaq_qMSdkf-zHUBWUtQT1k5aICgTUTpDN Enn9pDmnWQElEqMI5W1J4Y

257 Rebecca Falconer, "Pelosi announces war powers vote in attempt to limit Trump on Iran," Axios, Jan 5, 2020, https://www.axios.com/iran-crisis-pelosi-war-powers-vote-to-limit-trump-eeebd4a2-9874-40e8-b1a5-956fff39d922.html?utm_source=facebook&utm_medium=social&ut&fbclid=IwAR1DqMYTUl sx3-TCYbz39SJ0bZczJBTnNzOQP9pBwsGAgp-6JmKettXCHzo

258 Emma Tucker, "Trump to Congress: My Tweets Will Notify You of Military Action Against Iran," The Daily Beast, Jan. 05, 2020, https://www.thedailybeast.com/trump-tells-congress-his-tweets-serve-as-notification-of-military-action-against-iran?ref=home

259 T.C. Sottek, "Trump tells Congress to follow him on Twitter for updates on war with Iran," The Verge, Jan 5, 2020, https://www.theverge.com/2020/1/5/21050757/trump-iran-qassem-soleimani-attack-congress-twitter-follow-updates?fbclid=IwAR1k-bCHI92EViZNqgGTK4Wh Tm1Kd9kYX6sKBkLNp6exbRGf42-ut70eLmQ

260 Greg Sargent, "GOP senator who erupted over Iran briefing shares awful new details," Washington Post, Jan. 9, 2020, https://www.washingtonpost.com/opinions/2020/01/09/

gop-senator-who-erupted-over-iran-briefing-shares-awful-new-details/?fbclid=IwAR2qffkgQEu Nxrak1DM4TYT3VF0k7JQ3jIBgKyMOEoYMNbJm1UOxYumayWI

261 Oona A. Hathaway, "The Soleimani Strike Defied the U.S. Constitution," The Atlantic, January 4, 2020, https://www.theatlantic.com/ideas/archive/2020/01/soleimani-strike-law/604417/?fbclid=IwAR1l-9f56PublFDxlpjJum4svfYU2KJlNUcXJqkde9i4JRiUTeDtlOJhL Es

262 Jim Webb, "When did it become acceptable to kill a top leader of a country we aren't even at war with?" Washington Post, January 9, 2020, https://www.washingtonpost.com/opinions/ the-iran-crisis-isnt-a-failure-of-the-executive-branch-alone/2020/01/09/cc0f3728-3305-11ea-9313-6cba89b1b9fb_story.html

263 Max Boot, "William Barr's chilling defense of virtually unlimited presidential power," Washington Post, November 17, 2019, https://www.washingtonpost.com/ opinions/2019/11/17/william-barrs-chilling-defense-virtually-unlimited-presidential-power/

264 David Rohde, "William Barr, Trump's Sword and Shield," The New Yorker, January 13, 2020, https://www.newyorker.com/magazine/2020/01/20/william-barr-trumps-sword-and-shield?utm_campaign=aud-dev&utm_source=nl&utm_brand=tny&utm_mailing=TNY Magazine_Daily_011320&utm_medium=email&bxid=5be9e3713f92a40469fa2eae&cndid=5 3679941&esrc=&mbid=&utm_content=B&utm_term=TNY_Daily&verso=true

265 "Republicans Now Are More Open to the Idea of Expanding Presidential Power," Pew Research Cetner, August 7, 2019, https://www.people-press.org/2019/08/07/ republicans-now-are-more-open-to-the-idea-of-expanding-presidential-power/

266 United States Senate, "(U)Report of the Select Committee on Intelligence United States Senate on Russian Active Measures Campaigns and Interference in the 2016 U.S. Election," 11

267 Lee Moran," Carl Bernstein Delivers Ominous Assessment On America Ahead Of Impeachment Hearings," HuffPost, November 13, 2019, https://www.huffpost.com/entry/ carl-bernstein-donald-trump-impeachment-hearings_n_5dcbb87ee4b03a7e0291fb39?gucc ounter=1&guce_referrer=aHR0cHM6Ly93d3cuaHVmZnBvc3QuY29tLw&guce_referrer sig=AQAAAKwRUlnnUsiIp1Cwzby2U_mb0akibw4GeSPLbEUxrvwoI6qBj40Grmy3-TxB3v TfSjXAvBm6NCrAXcvFgw1IKji9E98pv0PYtxpx-sjFNfsXaDVuCydCgWOlT3I5nnkP4OWm dFfIndlPleE4QAuwlMS8Ucqub-dnX8t-9GCyWZZe

268 Dennis Prager, "America's Second Civil War," Dennisprager.com, Jan. 24, 2017, https://www. dennisprager.com/americas-second-civil-war/

269 Mary Papenfuss, "Legendary Journalist Bill Moyers Says He Fears For The Nation For The First Time In His Life," Huffington Post, November 11, 2019, https://www.huffpost.com/ entry/bill-moyers-impeachment-cnn-survival_n_5dc8cd83e4b02bf579426375?guccoun ter=1&guce_referrer=aHR0cHM6Ly93d3cuaHVmZnBvc3QuY29tLw&guce_referrer sig=AQAAAD3x5PMBf3B2ONkz_0wCR-orE4TKnMwhirs7dkYoQaiauoy5OZEe1fZERWI o9Y0Xr_U52IRyKBA3L2ODVIgZCWZCU-ciyTpL2E7ELMopPOurIGQ2VY6wl29J3I7yht 7B-6mx4JmYYV59VXMe5xOmZLIqQ20EYTNNc7uT9pHyREKd

270 Jennifer Rubin, "The Intelligence Committee's report is a triumph," Washington Post, Dec. 3, 2019, https://www.washingtonpost.com/opinions/2019/12/03/ an-intelligence-committee-report-with-pizzazz/

271 Joel Kotkin, "Is America About to Suffer Its Weimar Moment?" The Daily Beast, Dec. 31, 2019, https://www.thedailybeast.com/is-america-about-to-suffer-its-weimar-moment?via=news letter&source=DDMorning&fbclid=IwAR3axSuv0NBc7tY6H3DfAzwyb148MjRHH2Jaopm p5dmDLsvmcvg906JNdzs

272 Sara Boboltz, "John Roberts Warns Americans They're Taking Democracy 'For Granted'" Huffington Post, January 2, 2020, https://www.huffpost.com/entry/john-roberts-supreme-court-democracy_n_5e0e18d5e4b0843d360f3305?ncid=APPLENE WS00001

273 Astead W. Herndon, "'Nothing Less Than a Civil War': These White Voters on the Far Right See Doom Without Trump," New York Times, Dec. 29, 2019, https://www.nytimes. com/2019/12/28/us/politics/trump-2020-trumpstock.html?smtyp=cur&smid=fb-nytimes&fb clid=IwAR1JO7o50TEmxbO3QACg_-1LhYJEsjLVDvTQooaHcQF5lP023qxYpG80Vw

274 Jeb Lund, "'It's the Power That Does Something to Me'- Dispatches from the Saturday night Trump rally." Esquire, Feb. 20, 2017, https://www.esquire.com/news-politics/a53275/ trump-melbourne-florida-rally/

275 Eric Levitz, "GOP Lawmaker Plotted Insurrections to Establish Christian State," New York Magazine, Dec. 24, 2019, http://nymag.com/intelligencer/2019/12/matt-shea-christian-terrorism-washington-report-ammon-bundy.html

276 Oath Keepers, Twitter.com, Sept. 30, 2019, https://twitter.com/Oathkeepers/status/1178549790847590400

277 Thomas B. Edsall, "No Hate Left Behind-Lethal partisanship is taking us into dangerous territory." New York Times, March 13, 2019, https://www.nytimes.com/2019/03/13/opinion/hate-politics.html

278 Chris Cillizza, "14 key political trends from the 2018 exit polls," CNN , November 14, 2018, https://edition.cnn.com/2018/11/13/politics/2018-exit-polls/index.html

279 "Partisan Antipathy: More Intense, More Personal," Pew Research Center, October 10, 2019, https://www.people-press.org/2019/10/10/partisan-antipathy-more-intense-more-personal/

280 "The American Identity: Points of Pride, Conflicting Views, and a Distinct Culture," The Associated Press-NORC Center for Public Affairs Research, 2017, http://apnorc.org/projects/Pages/HTML%20Reports/points-of-pride-conflicting-views-and-a-distinct-culture.aspx

281 "Political Polarization in the American Public, Section 3: Political Polarization and Personal Life," Pew Research Cetner, June 12, 2014, https://www.people-press.org/2014/06/12/section-3-political-polarization-and-personal-life/

282 John Avlon, "Polarization is poisoning America. Here's an antidote," CNN , November 1, 2019, https://www.cnn.com/2019/10/30/opinions/fractured-states-of-america-polarization-is-killing-us-avlon/index.html?fbclid=IwAR1dPh9Fgm1xit_8j199ijsd8DjNptgoXLnter88IC2FWT7swYorj7LlbyA

283 Yoni Appelbaum, "How America Ends: A tectonic demographic shift is under way. Can the country hold together?" The Atlantic, December, 2019, https://www.theatlantic.com/magazine/archive/2019/12/how-america-ends/600757/

284 Noah Berlatsky, "Trump voters motivated by racism may be violating the Constitution. Can they be stopped?" NBC News, Jan. 27, 2020, https://www.nbcnews.com/think/opinion/trump-voters-motivated-racism-may-be-violating-constitution-can-they-ncna1110356

285 Ali Vitali, Kasie Hunt and Frank Thorp V, "Trump referred to Haiti and African nations as 'shithole' countries," NBC News Jan. 11, 2018, https://www.nbcnews.com/politics/white-house/trump-referred-haiti-african-countries-shithole-nations-n836946

286 Rebecca Solnit, "The American civil war didn't end. And Trump is a Confederate president" The Guardian, Nov. 4, 2018, https://www.theguardian.com/commentisfree/2018/nov/04/the-american-civil-war-didnt-end-and-trump-is-a-confederate-president?CMP=share_btn_fb&fbclid=IwAR1f5-bnR4NmsLNnOZ3-TeT0txP9pR3aWdyu8JY5EqYTYqYHWqeDDK_pGiI

287 Rebecca Leber, "Republicans Think Obama's Acting Like a King on Public Lands," The New Republic, February 20, 2015, https://newrepublic.com/article/121102/gop-reacts-obama-protecting-more-national-monuments

288 Jack Holmes, "Trump Lackey Lou Dobbs Says the Quiet Part Out Loud: Americans Have 'Obligations to The Leader'" Esquire, Jan. 7, 2020, https://www.esquire.com/news-politics/politics/a30428924/lou-dobbs-obligations-to-the-leader-trump/

289 Chauncey Devega, "Mental health professionals read Trump's letter: A study in "the psychotic mind" at work," Salon, Dec. 20, 2019, https://www.salon.com/2019/12/20/mental-health-professionals-read-trumps-letter-a-study-in-the-psychotic-mind-at-work

290 Tom King, "War is Peace, Freedom is Slavery, Ignorance is Strength," WSAU.com, August 14, 2018, https://wsau.com/blogs/tom-kings-blog/54/war-is-peace-freedom-is-slavery-ignorance-is-strength/

291 Tom O'Connor, "Russian Official Cancels U.S. Visit, Saying 'Second American Civil War' Is Underway," Newsweek, Jan 10, 2019, https://www.newsweek.com/russia-cancel-visit-american-civil-war-1287282

292 Jamie Seidel, "US civil war 'coming': Russia seizes on talk of US split — but did it help generate that in the first place?" News.com.au, Nov. 5, 2018, https://www.news.com.au/world/us-civil-war-coming-russia-seizes-on-talk-of-us-split-but-did-it-help-generate-that-in-the-first-place/news-story/261c2002bb9a0318206fbffb680396cc

293 Philip Ewing, "Here's How Russia Runs Its Disinformation Effort Against The 2018 Midterms" NPR, October 23, 2018, https://www.npr.org/2018/10/23/659545242/heres-how-russia-runs-its-disinformation-effort-against-the-2018-midterms

294 Yes, California, Wikipedia, https://en.m.wikipedia.org/wiki/Yes_California?fbclid=IwAR2282VMX0RNfDkmLNyUe6SL4NjbuqVN9YxZ8cEQi8ztvQjqtDQAyZ0Lgog

295 Whitney Webb, "Why a Shadowy Tech Firm With Ties to Israeli Intelligence Is Running Doomsday Election Simulations," MPN News, Jan. 4, 2020, https://www.mintpressnews.com/cybereason-israel-tech-firm-doomsday-election-simulations/263886/

296 Department of Defense, Defense Science Board "Resilient Military Systems and the Advanced Cyber Threat" 2013, p 13

297 "Trunews Ranked Number One on Content Platform Spreaker," Trunews.com, Nov. 23, 2016, https://www.trunews.com/article/trunews-ranked-number-one-on-content-platform-spreaker

298 Alex Woodward, "Right-wing pastor says Trump supporters will 'hunt down' Democrats when he leaves office," The Independent, October 24, 2019, https://www.yahoo.com/news/wing-pastor-says-trump-supporters-165700816.html

299 Matthew Rosenberg, Nicole Perlroth and David E. Sanger, "'Chaos Is the Point': Russian Hackers and Trolls Grow Stealthier in 2020," New York Times, Jan. 10, 2020, https://www.nytimes.com/2020/01/10/us/politics/russia-hacking-disinformation-election.html?smid=nytcore-ios-share&fbclid=IwAR2dhg9_BCByOixZuaaLnp2rptxWrWXZN8iN-r6FSIAnellc4q9RWNTC_DT0

300 Kurt Eichenwald, Twitter.com, Nov. 1, 2019 https://twitter.com/kurteichenwald/status/1190367840454533120?s=12&fbclid=IwAR1kNuUetEgP_qAKE30WM6yKds8b-WbVCW5akE3249dpry-5YD0ab-VZ7jA

301 Thomas Pepinsky, "Why the Impeachment Fight Is Even Scarier Than You Think," Politico, October 31, 2019, https://www.politico.com/magazine/story/2019/10/31/regime-cleavage-229895?fbclid=IwAR2Y-nJn2SPAnPhgkUnqY-f4tB4WRbCVqVg_5ahIdb1AZhT_rITeJ5GJhWo

302 Darren Byler, "I Researched Uighur Society in China for 8 Years and Watched How Technology Became a Trap," The National Interest, October 27, 2019. https://www.yahoo.com/news/researched-uighur-society-china-8-134500323.html

303 Stephen Vincent Benét, "The Devil and Daniel Webster," The Saturday Evening Post, 1936304   Ellen Nakashima, "NSA found a dangerous Microsoft software flaw and alerted the firm — rather than weaponizing it," Washington Post, Jan. 14, 2020, https://www.washingtonpost.com/national-security/nsa-found-a-dangerous-microsoft-software-flaw-and-alerted-the-firm--rather-than-weaponize-it/2020/01/14/f024c926-3679-11ea-bb7b-265f4554af6d_story.html

305 Jory Heckman, "CISA demands 'emergency action' from agencies on Windows vulnerability patch," Federal News Network, January 14, 2020, https://federalnewsnetwork.com/cybersecurity/2020/01/cisa-demands-immediate-and-emergency-action-from-agencies-to-patch-windows-vulnerability/

306 Breanne Deppisch, "DHS Was Finally Getting Serious About Cybersecurity. Then Came Trump." Politico, December 18, 2019, https://www.politico.com/news/magazine/2019/12/18/america-cybersecurity-homeland-security-trump-nielsen-070149

307 Ellen Nakashima, "U.S. Cybercom contemplates information warfare to counter Russian interference in 2020 election," Washington Post, Dec. 25, 2019, https://www.washingtonpost.com/national-security/us-cybercom-contemplates-information-warfare-to-counter-russian-interference-in-the-2020-election/2019/12/25/21bb246e-20e8-11ea-bed5-880264cc91a9_story.html

308 Eric White, "IG: OPM could put up a better defense against cyber attacks," Federal News Network, December 2, 2019, https://federalnewsnetwork.com/federal-newscast/2019/12/ig-opm-could-put-up-a-better-defense-against-cyber-attacks/

309 Megan Reiss, "Reducing Cyber Vulnerabilities in Weapons Systems: A New Priority," Lawfare Blog, October 19, 2018, https://www.lawfareblog.com/reducing-cyber-vulnerabilities-weapons-systems-new-priority

310 Bill Chappell, "Cyber Tests Showed 'Nearly All' New Pentagon Weapons Vulnerable To Attack, GAO Says," NPR, October 9, 2018, https://www.npr.org/2018/10/09/655880190/cyber-tests-showed-nearly-all-new-pentagon-weapons-vulnerable-to-attack-gao-says

311 David E. Sanger and William J. Broad, "New U.S. Weapons Systems Are a Hackers' Bonanza, Investigators Find," New York Times, Oct. 10, 2018, https://www.nytimes.com/2018/10/10/us/politics/hackers-pentagon-weapons-systems.html

312 Matthew Rosenberg, Nicole Perlroth and David E. Sanger, "'Chaos Is the Point': Russian Hackers and Trolls Grow Stealthier in 2020," New York Times, Jan. 10, 2020, https://www.nytimes.com/2020/01/10/us/politics/russia-hacking-disinformation-election.html?smid=nytcore-ios-share&fbclid=IwAR2dhg9_BCByOixZuaaLnp2rptxWrWXZN8iN-r6FSIAnellc4q9RWNTC_DT0

313 Testimony of Dr. William LaPlante, Senior Vice President, Mitre National Security Sector, Before the Cybersecurity Subcommittee of the Senate Armed Services Committee, March 26, 2019 https://www.armed-services.senate.gov/imo/media/doc/LaPlante_03-26-19.pdf

314 Dan Lohrmann, "2019: The Year Ransomware Targeted State & Local Governments," Government Technology, December 23, 2019, https://www.govtech.com/blogs/lohrmann-on-cybersecurity/2019-the-year-ransomware-targeted-state--local-governments.html

315 Dominique Mosbergen, "Pentagon Restricts Use Of Fitness Trackers, Location-Tracking Apps Over Security Concerns," HuffPost, August 7, 2018, https://www.huffpost.com/entry/pentagon-fitness-tracker-gps-apps_n_5b6939a2e4b0de86f4a49d21

316 David Ignatius, "Russians are masters of deception when it comes to cyberwars," Washington Post, Jan. 2, 2020, https://www.washingtonpost.com/opinions/russians-are-masters-of-deception-when-it-comes-to-cyberwars/2020/01/02/d931bb70-2db4-11ea-9b60-817cc18cf173_story.html

317 Nicole Perlroth and Matthew Rosenberg, "Russians Hacked Ukrainian Gas Company at Center of Impeachment," New York Times, Jan. 13, 2020, https://www.nytimes.com/2020/01/13/us/politics/russian-hackers-burisma-ukraine.html

318 Isabelle Khurshudyan, "Russia's government overhauled as Putin looks to cement influence after presidency," Washington Post, Jan. 15, 2020, https://www.washingtonpost.com/world/europe/putin-proposes-strengthening-parliament-even-while-keeping-his-own-powers-intact/2020/01/15/695eac6a-36e5-11ea-a1ff-c48c1d59a4a1_story.html

319 Guy Caspi, "Why Are We Losing The Cyberwar?" Forbes, Jan 22, 2020, https://www.forbes.com/sites/forbestechcouncil/2020/01/22/why-are-we-losing-the-cyberwar/#69669d86b80d

320 Cyberspace Solarium Commission Report, 2020, p 5

321 Zoe Schiffer, "Mark Zuckerberg on lies in political ads: 'I don't think it's right for a private company to censor politicians'" The Verge, Oct. 17, 2019, https://www.theverge.com/2019/10/17/20919223/mark-zuckerberg-facebook-speech-live-politics-threats-free-expression

322 "Senators Introduce Bipartisan Legislation to Develop American 5G Alternatives to Huawei," HomelandSecurityToday.us (Government Technology & Services Coalition), January 14, 2020, https://www.hstoday.us/subject-matter-areas/cybersecurity/senators-introduce-bipartisan-legislation-to-develop-american-5g-alternatives-to-huawei/

323 Zak Doffman, "Huawei: U.S. Senators Compare Company To KGB As They Question New Pentagon Decision," Forbes, Jan. 24, 2020, https://www.forbes.com/sites/zakdoffman/2020/01/24/us-senators-demand-huawei--explanation-from-pentagon-its-like-working-with-the-kgb/#5abafd23399d

324 Josh Gold, "A Multistakeholder Meeting at the United Nations Could Help States Develop Cyber Norms," Council on Foreign Relations, January 16, 2020, https://www.cfr.org/blog/multistakeholder-meeting-united-nations-could-help-states-develop-cyber-norms

325 Lesley Stahl, "The Great Brain Robbery- Economic espionage sponsored by the Chinese government is costing U.S. corporations hundreds of billions of dollars and more than two million jobs," 60 Minutes, Jan. 17, 2016, https://www.cbsnews.com/news/60-minutes-great-brain-robbery-china-cyber-espionage/

326 John Burtka, "Our overdue reckoning with China," Washington Post, Oct. 20, 2019, https://www.washingtonpost.com/opinions/2019/10/20/our-overdue-reckoning-with-china/